Muhamad Syazwan Faid is a senior lecturer at the Department of Islamic Studies, Centre for General Studies and Co-Curricular, Universiti Tun Hussein Onn Malaysia (UTHM). His expertise lies at the intersection of Islamic Astronomy and computational sciences, where he integrates traditional Islamic astronomical practices with modern technological advancements.

Muhammad Syaoqi Nahwandi is a lecturer at UIN Syekh Nurjati Cirebon, where he specializes in Islamic Astronomy and its integration with modern scientific techniques. His academic focus includes lunar crescent visibility, prayer time calculations, and Hijri calendar determination.

Shahrin Ahmad is a renowned Malaysian astronomer and the founder of Falak Online, a platform dedicated to promoting Islamic Astronomy and public astronomy education. With extensive experience in observational astronomy, Shahrin is widely regarded as a leading figure in advancing understanding and application of Islamic astronomical practices.

Mohd Saiful Anwar Mohd Nawawi is an associate professor at the Department of Fiqh-Usul and Applied Sciences, Academy of Islamic Studies, University of Malaya (UM). A renowned scholar in Islamic Astronomy, he specializes in the development of methodologies for Hijri calendar determination, Qibla direction, and prayer time calculations.

Python for Islamic Astronomy

Modern Computational Approaches to Hijri Calendar, Qibla, and Prayer Times

Muhamad Syazwan Faid,
Muhammad Syaoqi Nahwandi,
Shahrin Ahmad, and
Mohd Saiful Anwar Mohd Nawawi

CRC Press
Taylor & Francis Group
Boca Raton London New York

CRC Press is an imprint of the
Taylor & Francis Group, an informa business

Designed cover image: Shutterstock

First edition published 2026
by CRC Press
2385 NW Executive Center Drive, Suite 320, Boca Raton FL 33431

and by CRC Press
4 Park Square, Milton Park, Abingdon, Oxon, OX14 4RN

CRC Press is an imprint of Taylor & Francis Group, LLC

ISBN: 978-1-041-08261-3 (hbk)
ISBN: 978-1-041-09243-8 (pbk)
ISBN: 978-1-003-64912-0 (ebk)

DOI: 10.1201/9781003649120

Typeset in Times
by Deanta Global Publishing Services, Chennai, India

Contents

Acknowledgments

Alhamdulillah, praise be to Allah SWT for His Rahmah and with His permission we managed to complete this book successfully. With His permission, too, we gained the enthusiasm and spiritual strength needed in weathering the trials and tribulations on my journey. We would also like to extend our grateful thanks to the parties who have been immensely helpful and supportive throughout the time we worked on this book. Without their help, cooperation, and support, we may not have been able to come this far. Finally, to all parties directly or indirectly involved and who contributed to this study, we pray that you will be blessed by Allah SWT and bestowed with the best rewards. Thanks for everything.

This research was supported by Universiti Tun Hussein Onn Malaysia (UTHM) through Tier 1 vot (Q989).

Application of Islamic Astronomy for Muslim

1

This book is supplemented by an interactive coding website hosted on a Google Colab repository. Readers can explore the code used in the computation and calculation of Islamic Astronomy, observe how it is implemented within the Colab environment, and adjust the demonstrated code according to their own variables and real-life situations. Users do not need to rewrite the code manually from this book; instead, they can simply copy it directly from the Google Colab repository. Rewriting code often leads to mistakes, and the more mistakes made, the less enjoyable the exercise becomes; trust me, I speak from experience. The website contains 6 hyperlinks, which direct the user to the corresponding chapter. The website can be accessed on bit.ly/pythonforis lamicastronomy, or using the QR Code given in Figure 1.1.

From this link, user can edit the code. The edited code will be saved on the user's own copy of Google Colab and won't affect the original coding.

Islamic Astronomy, also known as *'ilm al-hay'ah*, refers to the branch of science developed and practiced within the Islamic world that deals with the study of celestial bodies and their movements, primarily for religious, calendrical, and navigational purposes (Ilyas, 1997). This area of study combines observational astronomy with Islamic principles, making it a vital area of study for Muslims. Islamic Astronomy has historically contributed to various fields, from timekeeping and calendar formation to navigation and the understanding of celestial events (Yusuf, 2010). Its importance is seen in the precise ways it influences Muslims' religious obligations and broader scientific knowledge.

One of the most practical applications of Islamic Astronomy is determining the exact times for the five daily prayers (*Salah*), which rely on the sun's position in the sky. This ensures that Muslims perform their prayers at the correct times each day (Abas et al., 2022). In Malaysia, the Department of Islamic Development Malaysia (JAKIM) publishes detailed prayer timetables every

DOI: 10.1201/9781003649120-1

FIGURE 1.1 QR code for Google Colab website.

year, calculated using astronomical algorithms and verified through observation. Tools like the Almanak Falak JAKIM allow users to refer prayer times specific to the zone (Huda et al., 2014). The Umm al-Qura calendar in Makkah uses a combination of astronomical data and historical tradition to set prayer times and official schedules, especially critical for the millions of pilgrims during Hajj and Umrah (Rubin, 2017). The Muslim Council of Britain provides detailed prayer timetables for different cities, incorporating high-latitude adjustments for places like Edinburgh, where sunlight patterns complicate standard calculations (Ali, 2015).

The Hijri calendar is a lunar calendar in which each month begins with the sighting of the new crescent moon (*hilal*). This determination is critical for establishing the dates of key Islamic events such as Ramadan, Eid al-Fitr, and Hajj (Ilyas, 1984). Moon sighting activities are regularly held across Muslim-majority countries. For example, in Indonesia, the Ministry of Religious Affairs coordinates nationwide *rukyah* observations, such as those conducted at the Pelabuhan Ratu Observatory, to verify the start of Ramadan and other significant months (Wahidi et al., 2019). The State Mufti Department manages official crescent moon sightings from sites like Bukit Agok Observatory, with live national broadcasts of Ramadan and Shawwal announcements (Mohd Nawawi et al., 2024). Morocco is known for using *hisab* (astronomical calculations) alongside *rukyah* to determine the Islamic months, often leading to different Ramadan start dates compared to neighboring countries (Lairgi, 2025).

Muslims must face the Kaaba in Makkah when performing prayers. Islamic Astronomy provides methods for determining the Qibla direction accurately from any point on Earth. Historically, this was done using instruments like the astrolabe, and today, with the help of digital compasses and satellite technology (Faid, Nahwandi, et al., 2022). In recent years, researchers undertook a measurement project for mosque prayer spaces to ensure they were accurately aligned with the Qibla based on updated astronomical and geospatial data (Yildirim et al., 2024). Al-Azhar Mosque, one of the oldest universities in the Islamic world, underwent a Qibla realignment in 1992 after new astronomical surveys corrected earlier orientation errors dating back centuries (Rabbat, 1996). Muslim communities often use smartphone apps (e.g., Muslim Pro) and Google Earth Qibla services, which apply astronomical algorithms to pinpoint the Kaaba's direction from anywhere in the country (Schumm, 2020).

Islamic teachings emphasize observing and responding to celestial events such as solar and lunar eclipses, which are occasions for special prayers. Islamic astronomers calculate and predict these events with precision, enabling communities to observe them properly (King, 1993). During the total lunar eclipse on July 27, 2018, Muslim communities worldwide, including in Saudi Arabia and Indonesia, performed *Salat al-Khusuf* (eclipse prayer) and organized observation sessions. Observatories like Bosscha Observatory in Indonesia broadcasted the event live, blending scientific outreach with religious observance (Izzuddin et al., 2022). During the 2019 annular solar eclipse, mosques across the kingdom held *Salat al-Kusuf*, and observatories provided live coverage with explanations linking religious practice to astronomical phenomena (Elmhamdi et al., 2024).

Before the advent of modern navigation tools, Islamic astronomers developed sophisticated navigation techniques based on celestial bodies. This was especially crucial for travelers, pilgrims, and maritime traders navigating long distances across deserts and seas. Muslim sailors and traders navigating the Indian Ocean between Africa, the Arabian Peninsula, and Southeast Asia used instruments like the *Kamal*, a simple device to measure the altitude of stars, to determine their position and direction. This knowledge was crucial in facilitating trade routes and pilgrimage journeys for centuries (Niri et al., 2023). North African caravans traveling across the Sahara to Makkah relied on astronomical observations, notably the Pole Star (Polaris), to maintain their route across the desert. During the Ottoman Empire, maritime explorers and military fleets used Islamic Astronomy-based charts and instruments like the astrolabe and quadrant to navigate the Mediterranean and Red Sea efficiently (Faid, Nawawi, et al., 2022).

2 Why Python Matters in Islamic Astronomy

Python is a widely used, interpreted, object-oriented, high-level programming language with dynamic semantics, designed for general-purpose programming. It is ubiquitous, and many people use Python-powered devices every day, whether they realize it or not. Python was created by Guido van Rossum and was first released on February 20, 1991. While you might know the python as a large snake, the name of the Python programming language actually comes from an old BBC television comedy sketch series called "Monty Python's Flying Circus" (Mehare et al., 2023). One of the great features of Python is that it is truly the work of one person. Typically, new programming languages are developed and published by large companies employing many professionals, and due to copyright regulations, it is very difficult to name any individuals involved in such projects. Python is an exception (Faid, Mohd Nawawi, et al., 2024).

Of course, Guido van Rossum did not develop and expand all the components of Python by himself. The rapid spread of Python across the globe is the result of the continuous work of thousands of (very often anonymous) programmers, testers, users (most of them not IT experts), and enthusiasts, but it must be said that the initial idea (the seed from which Python sprouted) came from one person – Guido (Ozgur et al., 2021). Python is maintained by the Python Software Foundation, an organization and community dedicated to developing, improving, expanding, and popularizing the Python language and its environment. There are billions of lines of code written in Python, which means nearly unlimited opportunities for code reuse and learning from well-designed examples. Moreover, there is a large and highly active Python community, always happy to help (Faid, Nawawi, Saadon, et al., 2023).

 DOI: 10.1201/9781003649120-2

Several factors make Python great for learning:

- It's easy to learn – The time required to learn Python is shorter than many other languages, meaning you can start real programming more quickly.
- It's easy to use for writing new software – Often, it's possible to write code faster when using Python.
- It's easy to obtain, install, and use – Python is free, open-source, and cross-platform; not all languages can boast that.

Programming skills prepare you for a career in almost any industry and are essential if you want to pursue more advanced and higher-paying software development and engineering roles. Python is a programming language that opens more doors than others. With solid Python knowledge, you can work in various jobs and industries. And the more you understand Python, the more you can do in the 21st century. Even if you don't need it for work, you'll find it useful to know (Blank & Deb, 2020).

Python is a great language for science, particularly in astronomy. Various packages like NumPy, SciPy, Scikit-Image, and Astropy (to name just a few) are all excellent examples of Python's suitability for astronomy, and there are many use cases. [NumPy, Astropy, and SciPy are fiscally sponsored projects by NumFOCUS; Scikit-Image is an affiliated project.] These tools make it easier to use Python in various astronomical projects.

For example, the European Southern Observatory (ESO), which operates the Very Large Telescope (VLT), offers data for download on their site. You can visit www.eso.org/UserPortal and create a username for their portal. If you're looking for data from the SPHERE instrument, you can download full datasets for any nearby stars with exoplanet or protoplanetary disks. It's a fantastic and engaging project for any Pythonista to reduce that data and make the hidden planets or disks visible amid the noise (Rhodes, 2011).

By using the tools offered by NumPy, SciPy, Astropy, Scikit-Image, and many more in combination, with a little patience and persistence, it's possible to analyze vast amounts of available astronomical data to produce some stunning results. Python plays a significant role in Islamic Astronomy, particularly in determining prayer times, the Qibla direction, and the beginning of the Hijri months. These aspects are central discussions in Islamic Astronomy.

PRAYER TIMES CALCULATION

The determination of prayer times involves using the position of the sun as a reference to measure the start and end times of each prayer. The position of the sun after its zenith is used to determine Zuhr, the sun's shadow for Asr, sunrise and sunset for Maghrib and Syuruk, while the sun's light below the horizon is used for Fajr and Isha (Faid et al., 2019). Python, with its libraries such as NumPy, SciPy, and Astropy, can be used to perform precise calculations of the sun's position, which is crucial for determining accurate prayer times (Faid et al., 2021).

QIBLA DIRECTION DETERMINATION

The determination of the Qibla direction requires accurate readings of latitude and longitude. Historically, Islamic scholars used the position of the sun and solar eclipses to determine these coordinates (Faid, Husien, et al., 2016). Today, GPS provides latitude and longitude, but surveyors and astronomers still require sun position data to obtain the azimuth direction. Python can be used to calculate the azimuth angle accurately, incorporating latitude, longitude, and the sun's position, which is vital for ensuring that the Qibla direction is correct (Amin, 2018).

DETERMINING THE BEGINNING OF THE HIJRI MONTHS

The beginning of an Islamic month is determined by the visibility of the crescent moon (Shariff et al., 2016). The moon's visibility depends on its position relative to the sun, and the calculations involved are more complex than those for the sun because of the moon's faster and more dynamic movement influenced by the gravitational forces of the sun and the Earth. Python, through the use of libraries like Astropy, allows for precise calculations of the moon's position, which is essential for determining the new moon and thus the start of the Hijri month (Muztaba et al., 2023).

3 Importance of Accuracy of Calculation

The precision in calculating the positions of the sun and moon directly impacts the determination of prayer times, the Qibla direction, and the visibility of the crescent moon. If these calculations are inaccurate, it could lead to incorrect prayer times, Qibla direction, or the start of the Islamic month. Therefore, it is crucial for those in authoritative positions to use accurate calculations. Public institutions like JAKIM, JUPEM, and the State Mufti Departments in Malaysia are responsible for producing accurate data, which the public relies on (Shariff et al., 2017). For example, the Qibla direction of mosques and suraus must be certified by official surveyors and the Mufti Department to ensure that the direction used by the local Muslim community is accurate. Similarly, the release of Prayer Time Tables and the annual Hijri Calendar involves precise calculations and is usually vetted by the State Falak Council before being published (Faid, Shariff, et al., 2016).

As the public, they are not burdened with performing these precise calculations ourselves. Instead, we are encouraged to rely on authoritative sources for accurate information. However, increasing public awareness and understanding of Islamic astronomical calculations can enhance appreciation for this knowledge and reduce confusion and the spread of misinformation (Faid et al., 2018). By using Python in Islamic Astronomy, we can harness its power to perform accurate and efficient calculations, ensuring that the Islamic practices related to time and direction are observed correctly and consistently (Faid et al., 2025). Accuracy in Islamic Astronomy is a highly critical aspect, particularly in determining the new Hijri month (Syazwan Faid et al., 2025). This is due to several factors that have a direct impact on the religious practices of Muslims, as well as broader social and economic implications (Wahyuni et al., 2022).

DOI: 10.1201/9781003649120-3

ACCURATE DETERMINATION OF RELIGIOUS PRACTICES

The determination of the new Hijri month is crucial as it sets the dates for key events in the Islamic calendar, such as the beginning of Ramadan, Eid al-Fitr, and Eid al-Adha. Accuracy in determining the Hijri month ensures that practices like fasting and the Eid prayer are performed on the correct days. Even a deviation as small as 0.5 degrees in the moon's position can lead to differences in date determination, potentially causing confusion within the community (Adegoke, 2013, 2017).

SOCIAL AND ECONOMIC IMPACT

Errors in determining the Hijri month can lead to misunderstandings among Muslims, especially in societies where religious dates play a significant role in social and economic planning. For instance, a change in the date of Eid can affect various aspects such as holiday planning, festival preparations, and related economic activities (Mohd Nawawi et al., 2024).

IMPORTANCE OF MODERN TECHNOLOGY AND CALCULATIONS

The use of modern technology, such as programming in Python, and advanced astronomical tools, allows for more accurate and consistent calculations in determining the Hijri month. These calculations involve various astronomical factors, such as the positions of the moon, sun, and Earth, as well as other factors like weather conditions and topography that can affect the visibility of the new moon (Junaidi, 2022). By employing advanced algorithms and precise mathematical calculations, the likelihood of errors in determining the Hijri month is significantly reduced. This accuracy is essential to ensure that Muslims worldwide can practice their faith with full confidence that the dates they observe are correct and valid (Gharaybeh, 2025).

COMPLEXITY OF CALCULATIONS

The process of calculating the new Hijri month is highly complex, involving hundreds of mathematical formulas and thousands of variables. Every aspect must be meticulously calculated to ensure accurate results. For example, factors such as the moon's altitude angle, the distance between the moon and the sun, and the times of sunset and moonset all play crucial roles in these calculations. Any error in these calculations can lead to incorrect date determination, which not only affects religious practices but can also cause confusion and division among Muslims (Faid, Shariff, et al., 2024; Meeus, 1991). Therefore, accuracy in Islamic Astronomy is not only necessary to maintain the correctness of religious practices and rituals but also to ensure social harmony and stability within the community (Faid, Nawawi, et al., 2024). By leveraging technology and precise scientific approaches, Muslims can observe their religious practices with greater confidence and alignment with the teachings of their faith. With the advancement of technology, the use of astronomical calculation software and programming algorithms has greatly facilitated this process, ensuring that each calculation is done accurately and efficiently. However, even with the aid of technology, expertise and in-depth knowledge of Islamic Astronomy are still required to ensure accuracy in determining the new Hijri month (Rasyid et al., 2024).

Python can be utilized to model and visualize astronomical data in a highly efficient and accurate manner. Below are some specific applications of Python in the field of Islamic Astronomy (Falak).

CALCULATION AND VISUALIZATION OF CRESCENT MOON POSITION

Python can be used to accurately calculate the position of the moon each night. By applying precise astronomical formulas, it is possible to determine the time and location where the crescent moon will be visible. Python also allows us to plot the moon's movement in the form of graphs or maps, making it easier for astronomers to predict and confirm the visibility of the crescent moon. This is crucial for determining the start of the Hijri month and Islamic celebrations like Ramadan and Shawal (Rasyid et al., 2023).

CALCULATION AND VISUALIZATION OF PRAYER TIMES

Python can be employed to calculate prayer times based on the sun's position. By inputting geographical location data, Python can calculate sunrise, solar noon, sunset, and other critical times needed to determine prayer times. Additionally, Python can plot the sun's position and the brightness of the sky throughout the day. This helps astronomers and the general public understand the changes in prayer times throughout the year, including variations due to seasonal changes and location (Razzak, 2024).

CALCULATION OF QIBLA DIRECTION

Python can be used to calculate the sun's shadow trajectory to determine the Qibla direction. By calculating the sun's position at a specific time, we can determine the direction of the shadow that will indicate the accurate Qibla direction. This is particularly useful for ensuring that the prayer direction is correct, especially in places where visual guidance might not be available. Python can generate maps or graphs showing the Qibla direction for various locations around the world (Asrin et al., 2018). By using Python, we can simplify and expedite calculation processes that would otherwise require a significant amount of time and effort if done manually. Python not only simplifies the calculation process but also ensures that the results are accurate and reliable. This makes Python an invaluable tool in the field of Islamic Astronomy (Falak) and Islamic Astronomy at large (Al-Rajab et al., 2023; Loucif et al., 2024).

4 Setting up Python for Astronomical Calculations

GETTING STARTED WITH PYTHON

Google Colaboratory, commonly known as Google Colab, is a free cloud-based platform provided by Google that allows users to write and execute Python code through the browser. It is especially popular among data scientists, educators, and researchers for its simplicity, power, and seamless integration with Google Drive. No installation is required, just a Google account.

With Google Colab, you can:

- Write and run Python code instantly.
- Use pre-installed scientific libraries like NumPy, Pandas, Matplotlib, and more.
- Collaborate with others in real time.
- Save and share notebooks easily through Google Drive.

This makes it a convenient tool for projects such as calculating Islamic Astronomy matter using Python.

DOI: 10.1201/9781003649120-4

USING PYTHON FOR PRAYER TIME CALCULATIONS IN GOOGLE COLAB

You can use Google Colab to run Python scripts that calculate prayer times based on astronomical data and geographical coordinates. Here's how to get started:

Step 1: Access Google Colab
Go to https://colab.research.google.com. You'll need to sign in with your Google account.
Step 2: Create a New Notebook
- Click on **"File" > "New Notebook"**.
- A new notebook with a Python environment will open in your browser.

Step 3: Install dependencies or libraries and start coding!

BASIS OF PYTHON OPERATION

Python can handle basic arithmetic just like a scientific calculator. Below are the key operations:

Addition

```
a = 10
b = 5
result = a + b
print(result) # Output: 15
```

Subtraction

```
a = 10
b = 5
result = a - b
print(result) # Output: 5
```

Multiplication

```
a = 10
b = 5
result = a * b
print(result) # Output: 50
```

Division

```
a = 10
b = 5
result = a / b
print(result) # Output: 2.0
```

Integer Division

```
a = 10
b = 3
result = a // b
print(result) # Output: 3
```

Modulus

```
a = 10
b = 3
result = a % b
print(result) # Output: 1
```

Exponentiation

```
a = 2
b = 3
result = a ** b
print(result) # Output: 8
```

To use trigonometric functions, we import the math module. Python's math functions work in **radians**, not degrees.

Importing Math Module

```
import math
```

Convert Degrees to Radians

```
degrees = 90
radians = math.radians(degrees)
print(radians) # Output: 1.5707963267948966
```

Trigonometric Functions

Sine (sin)

```
angle = 30
result = math.sin(math.radians(angle))
print(result) # Output: 0.49999999999999994
```

Cosine (cos)

```
angle = 60
result = math.cos(math.radians(angle))
print(result) # Output: 0.5000000000000001
```

Tangent (tan)

```
angle = 45
result = math.tan(math.radians(angle))
print(result) # Output: 0.9999999999999999
```

Inverse Tangent (Arctangent/atan)

The atan() function returns the angle (in **radians**) whose tangent is the given number. To convert it into degrees, use math.degrees():

```
x = 1 # tan(angle) = 1
angle_rad = math.atan(x)
angle_deg = math.degrees(angle_rad)
print(angle_rad) # Output: 0.7853981633974483
(radians)
print(angle_deg) # Output: 45.0 (degrees)
```

This is especially useful in triangle calculations, astronomy, and physics when determining angles based on known side ratios.

BASIC CONDITIONAL STATEMENTS IN PYTHON

Conditional statements allow your program to make decisions based on certain conditions. This is useful in nearly all practical applications.

The if Statement

The if statement executes a block of code only if a specified condition is true.

```
temperature = 30
if temperature > 25:
   print("It's a hot day.")
```

The else Statement

The else block runs when the if condition is false.

```
temperature = 20
if temperature > 25:
   print("It's a hot day.")
else:
   print("It's a nice day.")
```

Using or in Conditions

The or keyword checks if **at least one** of multiple conditions is true.

```
day = "Sunday"
is_holiday = True

if day == "Sunday" or is_holiday:
   print("You can rest today.")
else:
   print("It's a working day.")
```

Combined Example

Here is a real-world style example:

```
score = 80
if score >= 90:
   print("Excellent")
elif score >= 70:
   print("Good job")
else:
   print("Keep trying")
```

Explanation

- If the score is 90 or more → print **"Excellent"**
- If the score is 70 or more (but less than 90) → print **"Good job"**
- If the score is less than 70 → print **"Keep trying"**

Output

```
Good job
```

Another Example Using or

```
score = 60
if score < 70 or score > 100:
   print("Score needs review")
else:
   print("Score is acceptable")
```

Output

```
Score needs review
```

These basic control structures allow your Python programs to make decisions and react to different inputs. This shows how Python checks conditions in order. It stops at the first condition that is True. These if, else, and or statements give your Python program logic, allowing it to behave differently based on input or conditions. This is essential in automation, data filtering, simulations, and more.

print() in Python

The print() function is used to display information or output on the screen. It is one of the most basic and commonly used functions in Python programming.

```
print("Welcome to the world of Python!")
```

Output

```
Welcome to the world of Python!
```

You can also print numbers or the result of calculations:

```
print(5 + 3)
```

Output

```
8
```

The print() function is essential for checking values, debugging, and interacting with users. Sometimes, you want to print a sentence that includes variable values, such as a person's name or age. In Python, this is made easier and cleaner with formatted strings, also called f-strings.

```
name = "Aminah"
age = 18
print(f"My name is {name} and I am {age} years old.")
```

Output

```
My name is Aminah and I am 18 years old.
```

In this example, the f before the quotation mark tells Python to treat the string as a **formatted string**. Inside the curly braces {}, Python will insert the value of the variable. You can even include calculations:

```
length = 5
width = 3
print(f"The area is {length * width} square units.")
```

Output

```
The area is 15 square units.
```

Formatted strings make your code shorter, cleaner, and easier to understand, especially when combining text and values. Sometimes you want to control the number of decimal places shown. You can use format specifiers like :.2f to round to 2 decimal places.

```
azimuth_kiblat = 292.347182
print(f"Azimuth Kiblat: {azimuth_kiblat:.2f}")
```

Output

```
Azimuth Kiblat: 292.35
```

In {azimuth_kiblat:.2f}:

- : starts the format
- .2f means **2 decimal places** in **floating-point format**

This is very useful in math, geography, and science applications.

CONVERTING DECIMAL DEGREES TO DEGREES, MINUTES, AND SECONDS

In many fields such as geography, surveying, and navigation, angles are often represented not just in decimal degrees (e.g., 292.35°) but in a more traditional format: Degrees, Minutes, and Seconds (DMS). While decimal degrees are easier for calculations, DMS format is more readable and commonly used in GPS coordinates, navigation systems, and official land records.

Given a decimal degree value:

1. **Degrees**: The whole number part of the decimal degree
2. **Minutes**: Multiply the decimal part by 60, then take the whole number
3. **Seconds**: Multiply the remaining decimal by 60 again

Example: Convert 292.35° to DMS

1. **Degrees** = 292
2. **Decimal part** = 0.35
3. **Minutes** − 0.35 × 60 = 21.0 → **21 minutes**
4. **Seconds** = 0.0 × 60 = 0 → **0 seconds**

Final Answer

292.35∘=292∘ 21′ 0″292.35^\circ = 292^\circ\ 21'\ 0''292.35∘=292∘ 21′ 0″

Here is a Python script that performs this conversion:

```
decimal_degree = 292.35

# Step 1: Get whole degrees
degrees = int(decimal_degree)
# Step 2: Get the decimal part and convert to minutes
decimal_part = decimal_degree - degrees
minutes_total = decimal_part * 60
minutes = int(minutes_total)

# Step 3: Convert the decimal part of minutes to seconds
seconds = round((minutes_total - minutes) * 60)

print(f"{decimal_degree}° = {degrees}° {minutes}'
{seconds}"")
```

Output

```
292.35° = 292° 21' 0"
```

This method is useful for displaying locations and angles in a standardized, human-friendly format.

INSTALLING PYTHON PACKAGES WITH PIP

As you begin writing Python programs, you will often need to use additional tools, libraries, or frameworks that are not included in Python by default.

These tools are called packages. Python provides a powerful command-line tool called pip to help you install and manage these packages easily.

What Is pip?

pip stands for "Pip Installs Packages". It is the official package installer for Python.

With pip, you can:

- Install new Python libraries (like numpy, matplotlib, flask)
- Upgrade packages
- Uninstall packages
- Check what packages are installed

Installing a Package Using pip

You can install a Python package from the internet using this simple command:

```
pip install package_name
```

Example

To install requests, a popular package for HTTP:

```
pip install requests
```

Listing Installed Packages

To view all packages currently installed:

```
pip list
```

This will show the name and version of each package. pip is a tool for installing and managing Python packages. You can install, upgrade, and remove packages easily using simple commands. pip connects to the Python Package Index (PyPI) to download packages from the internet. Learning to use pip is essential for working with real-world Python projects.

WORKING WITH DATES AND TIMES IN PYTHON, THE DATETIME MODULE

In many real-world programs, especially those involving calendars, logs, astronomical events, or scheduling, you need to work with dates and times. Python makes this easy through a built-in module called datetime.

The datetime module allows you to:

- Create and manipulate dates and times
- Extract day, month, year, hour, minute, etc.
- Perform date arithmetic (e.g., add days, subtract dates)
- Format dates into readable strings

Importing the Datetime Module

To use the datetime features, you first need **to import it:**

```
from datetime import datetime
```

This line imports the datetime class from the datetime module.

Creating a Specific Date

You can define a date and time using the datetime() constructor:

```
date = datetime(2024, 1, 14)
```

This creates a datetime object representing:

- **Year**: 2024
- **Month**: January (1)
- **Day**: 14

By default, the time is set to midnight (00:00:00) if not specified.

You can also include the time:

```
date = datetime(2024, 1, 14, 15, 30)
```

This means 14 January 2024, at 3:30 PM.

Accessing Date Components

Once a datetime object is created, you can access parts of it:

```
print(date.year) # 2024
print(date.month) # 1
print(date.day) # 14
```

You can also get the **day of the week** or **day of the year**:

```
print(date.weekday()) # 6 (Sunday; Monday is 0)
print(date.timetuple().tm_yday) # 14 (14th day of the
year)
```

Why Is This Useful?

Handling dates and times accurately is essential in:

- Astronomical calculations (like Qibla direction)
- Logging timestamps in applications
- Scheduling tasks or sending reminders
- Managing calendar events

Loop

A **loop** in programming is used to repeat a set of instructions **if a certain condition is true**. Loops are useful when we want to:

- Repeat a task many times
- Search for a value
- Simulate continuous or time-based processes

Python provides two main types of loops:

- for loop – used when you know how many times you want to repeat
- while loop – used when you want to repeat **until** a condition is met

A while loop continues **executing** the code inside it **as long as** the condition is true.

```
while condition:
# do something
If the condition becomes false, the loop stops.
```

Example 1: Simple Counter

```
count = 0
while count < 5:
print("Counting:", count)
count += 1
Counting: 0
Counting: 1
Counting: 2
Counting: 3
Counting: 4
```

Why Use while?

Use while when:

- You don't know how many times the loop should run.
- You're waiting for a condition to become true.

Example

Write a loop that prints numbers until the square of the number is greater than 100.

python

```
n = 1
while True:
print(n)
if n * n > 100:
break
n += 1
```

Summary

- while loops are perfect for conditions with **unknown repetitions**.
- Always ensure there's a way to **exit the loop**.
- They're useful for simulations, astronomy calculations, and search-based problems.

USING SKYFIELD

Skyfield is a Python library designed to calculate the positions of stars, planets, and satellites orbiting the Earth. Its results are expected to align with the positions generated by the United States Naval Observatory and their Astronomical Almanac within 0.0005 arcseconds (half a milli-arcsecond, or mas). Skyfield is written entirely in pure Python and can be installed without any compilation, making it accessible for various Python environments. It supports Python versions 2.6, 2.7, and 3, with NumPy as its only binary dependency. NumPy is a fundamental package for scientific computing with Python, and its vector operations make Skyfield efficient. Before Skyfield was developed, an older version known as PyEphem was used. Skyfield builds on the capabilities of PyEphem, offering more accurate and modern astronomical calculations (Faid, Mohd Nawawi, et al., 2023; Faid, Nawawi, & Saadon, 2023; Faid, Nawawi, Wahab, et al., 2023).

PyEphem uses popular astronomical calculation techniques derived from Jean Meeus's *Astronomical Algorithms*, such as the IAU 1980 model of Earth's nutation and the VSOP87 planetary theory. These techniques are still employed by various authoritative bodies around the world (Holwerda et al., 2016). PyEphem offers accuracy up to 1 arcsecond, which is sufficient for calculating lunar and solar data. However, for new projects, the Skyfield library should be prioritized over PyEphem. Its modern design encourages better Python coding practices and utilizes NumPy to accelerate calculations. PyEphem's reliance on C libraries often results in frustrating installation issues. If the Python Package Index (PyPI) doesn't have a wheel for your system, you would need a C compiler and a Python development environment to install PyEphem.

Another drawback of PyEphem is its handling of angular units, which can be confusing. The library tries to be clever by interpreting string inputs like '1.23' as degrees of declination (or hours, when setting the right ascension), but floating-point inputs like 1.23 are assumed to be in radians. The angles returned by PyEphem add to the confusion: when printed, they display in degrees, but performing arithmetic on them reveals that they are in radians. This leads to significant confusion and makes the code harder to read, but fixing it would break existing programs that rely on PyEphem. Moreover, PyEphem's compute () method modifies its objects in place rather than returning a result. While this reflects how the underlying C library operates, it makes using compute () in list comprehensions difficult, you end up with a list of None objects. Therefore, in this context, Skyfield will be used for calculating lunar and solar data. Skyfield's advantages, such as modern design, better integration with Python, and ease of use, make it a superior choice for astronomical calculations in contemporary projects. Skyfield is a modern Python library that makes it easy to compute positions of planets and other celestial objects using high-precision data from NASA's Jet Propulsion Laboratory (JPL). This guide explains how to install and use the **DE440s ephemeris**, which contains positional data for the solar system.

Step 1: Install the Skyfield Library

To begin, install the Skyfield library in Google Colab or Jupyter Notebook. In a new cell, type the following and press **Shift + Enter**:

```
!pip install skyfield
```

This will install the core Skyfield library, which is needed to perform astronomical computations.

Step 2: Understand What an Ephemeris Is

An **ephemeris** is a table or dataset that provides the positions of celestial objects at regular intervals. Skyfield uses ephemerides published by NASA JPL to calculate accurate positions of planets and the moon. For this example, we will use **DE440s**, a compact but high-accuracy ephemeris suitable for many applications.

Step 3: Load the Ephemeris DE440s

Next, we will load the DE440s ephemeris using Skyfield. Type the following code into a new cell:

```
from skyfield.api import load
# Load the time scale
ts = load.timescale()
# Load the ephemeris data (DE440s) from the internet
planets = load('de440s.bsp')
```

load.timescale() prepares the time system used in calculations. load('de440s.b sp') downloads and loads the DE440s ephemeris file. This file will be downloaded from the internet the first time you run the code. Make sure your internet connection is stable.

Step 4: Verify Successful Loading

Once the ephemeris file is downloaded, Skyfield will store it locally (or in the Colab session temporarily). If successful, no error will be shown, and you can now access planetary positions like this:

```
earth = planets['earth']
mars = planets['mars']
```

Qiblah Calculation Using Python 5

The calculation of the Qibla direction is based on three points of location. The first is the location of the Kaaba, which is located at 21.4225 North, 39.8262 East. The second location is the calculated position of the observer. The third point is the north pole, where it acts as the axis for the angle between Kaaba and the calculated position. Therefore, the formula for the Qibla direction toward the Kaaba from the calculated position is,

$$\Delta\lambda = \left(\lambda_{Location} - \lambda_{Kaabah}\right) \quad \text{Equation 5.1}$$

$\Delta\lambda$ is the formula for calculating the distance from the longitude of the Kaaba to the longitude of the place for a location located in the east longitude, which can be obtained by calculating the difference in the values of the two longitudes. Then,

$$\mathrm{A} = \sin \Delta\lambda \quad \text{Equation 5.2}$$

$$\mathrm{B} = \cos\left(\varphi_{Location}\right) \times \tan(\varphi_{Kaabah}) \quad \text{Equation 5.3}$$

$$\mathrm{C} = \sin \varphi_{Location} \times \cos \Delta\lambda \quad \text{Equation 5.4}$$

$$\mathrm{D} = A / (B - C) \quad \text{Equation 5.5}$$

Then the Qibla Direction (θ) is calculated using the formula,

$$\theta = Tan^{-1}\left(D\right) \quad \text{Equation 5.6}$$

DOI: 10.1201/9781003649120-5

The formula for Qibla Calculation above can only be used for locations in Southeast Asia. This is because the calculation formula for calculating (Δλ) only considers conditions when a location is in the east longitude which has a value greater than the longitude of the Kaaba. To make the formula for calculating the direction of the Qibla can be used in all locations in the east and west longitudes, the distance from the longitude of the Kaaba to the longitude of the place (Δλ) is calculated using the following formula:

$$dev = abs\left(\lambda_{Location} - \lambda_{Kaabah}\right) \qquad \text{Equation 5.7}$$

$$\text{Distance between Longitude}, \Delta\lambda = \begin{cases} 360 - dev, if\ dev > 180 \\ dev, if\ dev \leq 180 \end{cases} \qquad \text{Equation 5.8}$$

The value of location longitude in the formula is positive if the location is in east longitude, and it is negative if the location is in west longitude.

In the Universal Qibla Direction Calculation, the reference value of the Qibla direction (θ) is not always from North to West. However, it can vary depending on the Qibla direction value (θ) and the Longitude of the Place ($\lambda_{location}$). To find out the reference of Qibla direction, use the following provisions:

$$\text{Arah Mata Angin Kiblat} = \begin{cases} UB & if\ \theta > 0\ and\ \lambda_{location} > \lambda_{Kaabah} \\ UT & if\ \theta > 0\ and\ \lambda_{location} \leq \lambda_{Kaabah} \\ UB & if\ \theta > 0\ and\ \lambda_{location} < 0\ and\ c \geq 180 \\ UT & if\ \theta > 0\ and\ \lambda_{location} < 0\ and\ c < 180 \\ SB & if\ \theta < 0\ and\ \lambda_{location} > \lambda_{Kaabah} \\ ST & if\ \theta < 0\ and\ \lambda_{location} \leq \lambda_{Kaabah} \\ SB & if\ \theta < 0\ and\ \lambda_{location} < 0\ and\ c \geq 180 \\ ST & if\ \theta < 0\ and\ \lambda_{location} < 0\ and\ c < 180 \end{cases}$$

"UB" is a statement that the Qibla direction uses a reference from the north to the west.

"SB" is a statement that the Qibla direction uses a reference from the south to the west.

"UT" is a statement that the Qibla direction uses a reference from the north to the east.

"ST" is a statement that the Qibla direction uses a reference from the south to the east.

The Azimuth of Qibla formula also varies depending on the reference of Qibla Direction value (θ). To find out the Azimuth of Qibla value, use the following provisions:

$$\text{Azimuth Kiblat} = \begin{cases} 360-\theta & \text{if } UB \\ 180-\theta & \text{if } SB \\ \theta & \text{if } UT \\ 180+\theta & \text{if } ST \end{cases}$$

QIBLA CALCULATION

Exercise 1

Determine the Qibla direction of Penang, which has geographical latitude of 5.2632 North and 100.4846 East. The Qaabah geographical latitude is 21.4225 North, longitude is 39.8262 East. First, determine the variables, which are written in Python as,

```
φ_Location = 5.2632
λ_Location = 100.4846
φ_Kaabah = 21.4225
λ_Kaabah = 39.8262
Difference_Longitude = (λ_Location-λ_Kaabah )
```

Second, perform the calculation. Using the above formula, the calculation can be written in Python as

In Python, this is written as

```
import math
A = math.sin(math.radians(abs(Difference_Longitude)))
```

To confirm the result of A, write in Python

```
print(A)
0.8717137230643722
```

Next, to calculate B to D, in Python is written as,

```
B = math.cos(math.radians(φ_Location))*math.tan(math.r
adians(φ_Kaabah))
```

```
print(B)
0.3906945822201198
C = math.sin(math.radians(φ_Location))*math.cos(math.r
adians(Difference_Longitude))
print(C)
0.0449496276198435
D = A/((B-C))
print(D)
2.5212623104570837
```

The direction of the Qibla θ is calculated in Python as

```
θ = math.degrees(math.atan(D))
print(θ)
68.36540021170762
```

The direction of the Qibla does not necessarily translate to the azimuth of the Qibla based on the magnetic compass. Our phone determines the azimuth based on the reference of the magnetic compass. Therefore, a minor computation needs to be added to determine the azimuth of the Qibla. First, correction of the difference in longitude based on Equation 5.8, written in Python as,

```
#Determine the Azimuth of the Qibla
if Difference_Longitude > 180:
  delta_λ = 360 - Difference_Longitude
else:
  delta_λ = Difference_Longitude
print(delta_λ )
60.6584
```

Then, the quadrant of the calculated user geographical location, with Kaaba as the center of the quadrant, in Python,

```
if θ > 0:
       if λ_Location > λ_Kaabah:
               quadrant = "UB" # Utara Barat
       elif λ_Location <= λ_Kaabah:
                quadrant = "UT" # Utara Timur
   elif λ_Location < 0:
               if c >= 180:
                               quadrant = "UB"
               else:
                               quadrant = "UT"
```

```
elif θ < 0:
   if λ_Location > λ_Kaabah:
      quadrant = "SB" # Selatan Barat
   elif λ_Location <= λ_Kaabah:
             quadrant = "ST" # Selatan Timur
elif λ_Location < 0:
    if c >= 180:
            quadrant = "SB"
            else:
                    quadrant = "ST"
print(quadrant)
UB
```

If the quadrant is UB, meaning Utara-Barat in Malay or Northwest in English. This means that the Azimuth of the Qibla is located at the Northwest from the computed location. To convert from Qibla direction to Qibla azimuth, written in Python

```
if quadrant == "UB":
  azimuth_kiblat = 360 - θ
elif quadrant == "SB":
  azimuth_kiblat = 180 - θ
elif quadrant == "UT":
  azimuth_kiblat = θ
elif quadrant == "ST":
  azimuth_kiblat = 180 + θ
print(azimuth_kiblat)
291.63459978829235
```

Therefore, the azimuth of the Qibla is 291.63 degree. To convert the result in degree,

```
# To Convert in Degree Form
degrees = int(azimuth_kiblat)
decimal_part = azimuth_kiblat - degrees
minutes_total = decimal_part * 60
minutes = int(minutes_total)
# Step 3: Convert the decimal part of minutes to
seconds
seconds = round((minutes_total - minutes) * 60)
print(f"{azimuth_kiblat}° = {degrees}° {minutes}'
{seconds}"")
291.63459978829235° = 291° 38' 5"
```

Finally, to print the result in a text

```
print(f'The azimuth of the Qibla for Location with
coordinates {φ_Location} N Latitude, {λ_Location}
Longitude, is {azimuth_kiblat:0.2f}')
The azimuth of the Qibla for Location with coordinates
5.2632 Latitude, 100.4846 Longitude, is 291.63
```

Exercise 2

Determine the Qibla direction of a location that has geographical latitude of 1.21 South and 108.411 East. Use the coding from the Google Colab website to generate the azimuth.

Exercise 3

Determine the Qibla direction for the city of Edinburgh, which has geographical latitude of 55.9533° North and 3.1883° West. This exercise is not available on the Google Colab website, meaning you need to make your own, based on the example provided.

Exercise 4

Calculate the azimuth of the Qibla for the city of Apia in Samoa with geographical latitude of 13.833 South and geographical longitude of 171.75 West.

Exercise 5

Calculate the azimuth of the Qibla for the city of Washington, DC, in the USA with geographical latitude of 38.904722 North and geographical longitude of 77.016389 West.

6 Istiwa' A'zam and Rashdul Qibla

Istiwa' A'zam refers to the astronomical event when the sun is directly above the Kaaba in Makkah. During this moment, the shadow of any vertical object anywhere on Earth points directly away from the Kaaba, making it one of the most accurate methods for determining the Qibla (prayer direction). This phenomenon is not affected by Earth's magnetic field, unlike compass-based Qibla methods, making it especially reliable.

Istiwa' A'zam happens twice a year, when the apparent path of the sun crosses the exact latitude of the Kaaba, which occurs on 28th May and 15th July approximately. On these dates, the sun passes directly overhead the Kaaba, around 12:16 p.m. (Saudi time).

At the time of Istiwa' A'zam: The sun is directly over the Kaaba. A vertical stick or object will cast a shadow that points exactly in the opposite direction of the Qibla. This method is free from magnetic disturbances, unlike a compass. Thus, shadows on these dates provide a natural, accurate Qibla direction, especially for mosques or homes without modern tools.

However, since Istiwa' A'zam only occurs twice a year, it is not always practical for everyday or long-term Qibla alignment. For this reason, Muslims use Rashdul Qibla, which refers to the general calculation of the Qibla direction using sun shadow. Rashdul Qibla refers to the determination of the Qibla direction using the sun's position and the direction of its shadow. When the sun's azimuth matches the azimuth of the Qibla, the shadow of any upright object will point directly opposite the Kaaba, effectively giving an accurate Qibla direction. This moment can be calculated for any location in the world, not just during the special events of Istiwa' A'zam (when the sun is directly above the Kaaba). Thus, Rashdul Qibla can be used on many dates throughout the year.

Modern tools like mobile compasses often suffer from magnetic interference, leading to significant inaccuracies, up to 20° in some cases. This could result in the prayer direction being off by over 1000 km from the Kaaba. In contrast, the sun's position can be precisely calculated using astronomical algorithms and ephemeris data. This makes Rashdul Qibla:

DOI: 10.1201/9781003649120-6

- Independent of magnetic errors.
- Applicable worldwide.
- Simple to verify using natural observation.

While Istiwa' A'zam only happens twice a year (around 28 May and 15 July), Rashdul Qibla can be calculated for any day using the method of finding when the sun's azimuth equals the Qibla azimuth for a given location. The position of the sun can be accurately calculated due to constant observation by the astronomer and improvement in ephemeris calculation by the astrometric researcher. The calculation of the time when the sun shadow is facing the Qibla is as follows. First step, calculate the first auxiliary angle to get the value of the sun's hour angle with the following formula:

$$\mathrm{U} = \tan^{-1}\left(\frac{1}{\tan - \theta \mathrm{x} \sin\varphi_{Location}}\right) \quad \text{Equation 6.1}$$

then calculate the second auxiliary angle with the following formula

$$\mathrm{T} - \mathrm{U} = \cos^{-1}\left(\frac{\tan\delta x \cos U}{\tan\varphi_{Location}}\right) \quad \text{Equation 6.2}$$

After the two auxiliary angles are obtained, then calculate the sun's hour angle using the following provisions:

$$t\_HMS) = \left(\begin{cases} -(T-U)+U, if\, U>0 \\ |T-U|+U, if\, U \le 0 \end{cases}\right) / 15 \quad \text{Equation 6.3}$$

add the sun's hour angle (in hours) by 12 o'clock to get the solar time of Rashdul Qibla:

$$R = 12 + t\_HMS \quad \text{Equation 6.4}$$

Finally, the Rashdul Qibla time according to local mean time can be obtained using the following formula:

$$Rashd\, alQibla\, Mean\, time = \begin{cases} R - EoT + \dfrac{TZ \times 15 - \lambda_{location}}{15}, if\, TZ > 0 \\ R - EoT - \dfrac{|TZ \times 15| - |\lambda_{location}|}{15}, if\, TZ \le 0 \end{cases}$$

Equation 6.5

Where θ is the Qibla direction, as calculated from the previous chapter. TZ is Time Zone of the location. δ is sun declination. EoT is Equation of Time. The formula of Equation of Time is complex and time consuming.

EQUATION OF TIME

The Equation of Time explains the difference between:

- Solar Time – time based on the sun's actual position in the sky (what a sundial shows)
- Clock Time – time shown by regular clocks (mean time)

This difference can be up to ±16 minutes. It changes every day of the year.

- Solar noon (when the sun is at its highest point) is not always exactly at 12:00 PM.
- The Equation of Time helps explain why the sun sometimes appears early or late.

TWO CAUSES OF THE DIFFERENCE

1. Earth's orbit is elliptical
    - Earth moves faster when close to the sun (January)
    - Earth moves slower when farther (July)
2. Earth's axis is tilted
    - The tilt (23.44°) changes the sun's path across the sky throughout the year

WHY DOES IT MATTER?

- Astronomy – correcting solar positions
- Sundials – to adjust for real sun time

- Islamic Prayer Times – calculating accurate Zuhur and other solar-based times
- Navigation – aligning maps and sun angles

SIMPLE DEFINITION

Equation of Time is equal to Solar time minus clock time. If EoT is positive, the sun is behind the clock. If EoT is negative, the sun is ahead of the clock

CALCULATION EQUATION OF TIME

There are several Python libraries that are able to calculate the Equation of Time. An example is pvlib. Pvlib requires installation from pip.

```
pip install pvlib
```

Once installed, pvlib only requires information of the day of the year. For example, to calculate the Equation of Time on January 14, 2024.

```
import pvlib
from datetime import datetime
import math
# define the date
date = datetime(2024, 1, 14)
# calculate the day of the year
day_of_year = date.timetuple().tm_yday
```

Then, to determine the result of the Equation of Time,

```
EoT = pvlib.solarposition.equation_of_time_pvcdrom(
day_of_year) / 60
print(f"Equation of Time on date {date} is {EoT}")
Equation of Time on date 2024-01-14 00:00:00 is
-0.14995874998914036
```

This Equation of Time is in decimal format, to convert as hour, minute, second.

```
# To Convert in Degree Form
degrees = int(EoT)
```

```
decimal_part = EoT - degrees
minutes_total = decimal_part * 60
minutes = int(minutes_total)
# Step 3: Convert the decimal part of minutes to
seconds
seconds = round((minutes_total - minutes) * 60)
print(f"{EoT}° = {degrees}° {minutes}' {seconds}"")
-0.14995874998914036° = 0° -8' -60"
```

The result is negative 8 minutes and 60 seconds. This means that the sun position is located ahead of the clock.

Exercise 1

Determine the Equation of Time for the date 26 May 2025.

Install the required libraries.

```
pip install pvlib
```

Input the date into Python variables.

```
Year = 2025
Month = 5
Day = 26
```

Input the date into the pvlib library

```
import pvlib
from datetime import datetime
import math
# define the date
date = datetime(Year, Month, Day)
# calculate the day of the year
day_of_year = date.timetuple().tm_yday
```

Input compute the EoT.

```
        EoT = pvlib.solarposition.equation_of_tim
        e_pvcdrom(day_of_year) / 60
print(f"Equation of Time on date {date} is {EoT}")
Equation of Time on date 2025-05-26 00:00:00 is
0.0519488998086564З
```

Convert to hour.

```
# To Convert in Degree Form
degrees = int(EoT)
decimal_part = EoT - degrees
minutes_total = decimal_part * 60
minutes = int(minutes_total)
# Step 3: Convert the decimal part of minutes to
seconds
seconds = round((minutes_total - minutes) * 60)
print(f"{EoT}° = {degrees}° {minutes}' {seconds}"")
0.05194889980865643° = 0° 3' 7"
```

The Equation of Time during 26 May 2025 is 3 minutes 7 second positive.

SUN DECLINATION

The sun declination can be calculated using the Skyfield Library. First, the Skyfield library needs to be installed.

```
pip install skyfield
```

Skyfield requires some information to operate; it requires general import of which Skyfield function the user want to use; the ephemeris, observation object, location of the user. First, the general import of the Skyfield function. For this case, we want to use the *load* function, to load ephemeris data and *N, S, W, E* to load geographical direction data, and *wgs84* to load earth geographical location data. In Python this is written as,

```
from skyfield.api import load
from skyfield.api import N, S, E, W, wgs84
```

Next, from the *load* function, determine which ephemeris and target that is going to be used. In this case, ephemeris 440s is used due to its small size and accuracy. Observation target is earth and sun, since this calculation only involves these two objects. In Python this is written as

```
ts = load.timescale()
eph = load('de440s.bsp')
planets = load('de440s.bsp')
```

```
earth = planets['earth']
sun = planets['sun']
```

Then, Skyfield requires location input to calculate the sun position. Let's say we calculate from the position of Kuala Lumpur, Malaysia, with coordinates of 3.1319° N, 101.6841° E, during the date of 14 October 2025.

```
name_city = "Kuala Lumpur"
lat_location = 3.1319
long_location = 101.6841
year = 2025
month = 10
day = 14
```

Next is to give the information to Skyfield user location. In this case, the user is from earth and calculation is computed on lat_location and long_location coordinates.

```
kuala_lumpur = earth + wgs84.latlon(lat_location,
long_location, elevation_m=0)
```

Then, input the date of the calculated observation, from the computed position to the targeted position. The computed position is kuala_lumpur, while the targeted position is sun. The code, written in Python as

```
astro = kuala_lumpur.at(ts.utc(year, month, day)).
observe(sun)
```

astro means that the input is in astrometric position of the sun. To convert into apparent position, as observed by the observer at kuala_lumpur, is;

```
app = astro.apparent()
```

From this app data, we can calculate the right ascension, declination, and distance of the sun, as observed from earth. To determine the declination of the sun using Skyfield.

```
app = astro.apparent()
ra_dec,dec_app,d_app = app.radec()
print(dec_app)
-08deg 01' 18.7"
```

Declination of the sun at Kuala Lumpur on 14 October 2025 is –08° 01' 18.7".

Exercise 2

Determine sun declination at Cape Town South Africa, during 26 May 2025, where it has a time zone of UTC/GMT +2 hours.

Install skyfield library

```
pip install skyfield
```

Import Necessary Function

```
from skyfield.api import load
from skyfield.api import N, S, E, W, wgs84
```

Load Ephemeris Data and Planet Objects

```
ts = load.timescale()
eph = load('de440s.bsp')
planets = load('de440s.bsp')
earth = planets['earth']
sun = planets['sun']
```

Insert the variable

```
#insert the variable
name_city = "Cape Town"
lat_location = -33.9221
long_location = 18.4231
year = 2025
month = 5
day = 26
tz = 2
```

Determine the sun declination

```
kuala_lumpur = earth + wgs84.latlon(lat_location,
long_location, elevation_m=0)
astro = kuala_lumpur.at(ts.utc(year, month, day)).
observe(sun)
app = astro.apparent()
app = astro.apparent()
ra_dec,dec_app,d_app = app.radec()
print(dec_app)
+21deg 04' 00.3"
```

Declination of the sun at Cape Town South Africa, on 26 May 2025 is +21° 04' 00.3".

CALCULATION OF RASHDUL KIBLAT

The calculation of Rashdul Kiblat involve calculating the Equation of Time and Sun Declination and the formula from 9 to 13. Using the previous example, on 26 May 2025, at Cape Town, where the Equation of Time is 0° 3′ 7″ and the sun declination is +21° 04' 00.3", we can determine the time when the sun shadow facing the Qibla direction.

First, determine the Qibla direction from the previous chapter coding.

Qibla Direction Coding

```
# Qibla Direction Calculation
φ_Location = -33.9221
λ_Location = 18.4231
φ_Kaabah = 21.4225
λ_Kaabah = 39.8262
Difference_Longitude = abs(λ_Location-λ_Kaabah )

#Calculation of Qibla Direction
import math
A = math.sin(math.radians(abs(Difference_Longitude)))
B = math.cos(math.radians(φ_Location))*math.tan(math.r
    adians(φ_Kaabah))
C = math.sin(math.radians(φ_Location)) * math.cos(m
    ath.radians(Difference_Longitude))
D = A/(B-C)
θ = math.degrees(math.atan(D))

#Determine the Azimuth of the Qibla
if Difference_Longitude > 180:
    delta_λ = 360 - Difference_Longitude
else:
    delta_λ = Difference_Longitude

if θ > 0:
   if λ_Location > λ_Kaabah:
       quadrant = "UB" # Utara Barat
   elif λ_Location <= λ_Kaabah:
       quadrant = "UT" # Utara Timur
   elif λ_Location < 0:
       if c >= 180:
           quadrant = "UB"
       else:
           quadrant = "UT"
```

```
elif θ < 0:
   if λ_Location > λ_Kaabah:
       quadrant = "SB" # Selatan Barat
   elif λ_Location <= λ_Kaabah:
        quadrant = "ST" # Selatan Timur
   elif λ_Location < 0:
       if c >= 180:
           quadrant = "SB"
       else:
           quadrant = "ST"

if quadrant == "UB":
   azimuth_kiblat = 360 - θ
elif quadrant == "SB":
   azimuth_kiblat = 180 - θ
elif quadrant == "UT":
   azimuth_kiblat = θ
elif quadrant == "ST":
   azimuth_kiblat = 180 + θ
print(azimuth_kiblat)
   23.354225930229862
```

The Qibla azimuth in Cape Town is 23.354225930229862.

To calculate Rashdul Qibla, from formula 9, determining U using Python code is written as

```
U = math.degrees(math.atan(1 / (math.tan(math.
radians(-azimuth_kiblat)) * math.sin(math.
radians(φ_Location)))))
print(U)
76.45187800936165
```

Sun declination, from previous coding is extracted from the variable dec_app. This is in degree format and cannot be included in computation. To express the sun declination in decimal format and make it computable, the code is

```
dec_app.degrees
```

Next, the value of T_U can be calculated using Equation 6.2 from sun declination and U variable. In Python this is written as,

```
T_U = math.degrees(math.acos(math.tan(math.radians(
dec_app.degrees)) * math.cos(math.radians(U)) / math.t
an(math.radians(lat_location))))
print(T_U)
97.71100015734693
```

After the T_U value is obtained, calculate the sun's hour angle (t_{HMS}) using Equation 6.3 using Python;

```
if U > 0:
   t_HMS = (-abs(T_U) + U) / 15
else:
   t_HMS = (abs(T_U) + U) / 15
print(t_HMS)
-1.4172748098656853
```

The Rashdul Qibla time, R in solar time, is obtained using Equation 6.4, and is expressed in Python as

```
R = 12 + t_HMS
print(R)
10.582725190134315
```

The value of R is in solar time. To change the direction of Qibla in solar time to local time, use Equation 6.5. Equation 6.5 requires data from time zone, I, and Equation of Time, EoT. In Python this is written as,

```
if tz > 0:
    Rashdul_Kiblat = R - EoT + (tz * 15 - long_
    location) / 15
else:
   Rashdul_Kiblat = R - EoT + (abs(tz * 15) -
   abs(long_location)) / 15
   print(Rashdul_Kiblat)
   11.302569623658991
```

The answer is in decimal. To convert to hour minute format,

```
# To Convert in Degree Form
degrees = int(Rashdul_Kiblat)
decimal_part = Rashdul_Kiblat - degrees
minutes_total = decimal_part * 60
minutes = int(minutes_total)
# Step 3: Convert the decimal part of minutes to
seconds
seconds = round((minutes_total - minutes) * 60)
print(f"{degrees}° {minutes}' {seconds}"")
11° 18' 9"
```

Rashdul Kiblat will occur, or The shadow of the sun, will point toward Qibla direction at 11:18 AM on 25 May 2025. At Cape Town South Africa.

Exercise 3

Determine the time of Rashdul Kiblat of Oslo in Norway, which has geographical latitude of 59.913333 North and 10.738889 East on July 30, 2025, that has Time zone UTC+1. The Qaabah geographical latitude is 21.4225 Noth, longitude is 39.8262 East.

Determine the Qibla direction

```
# Qibla Direction Calculation
φ_Location = 59.913333
λ_Location = 10.738889
φ_Kaabah = 21.4225
λ_Kaabah = 39.8262
Difference_Longitude = abs(λ_Location-λ_Kaabah )

#Calculation of Qibla Direction
import math

A = math.sin(math.radians(abs(Difference_Longitude)))
B = math.cos(math.radians(φ_Location))*math.tan(math.r
    adians(φ_Kaabah))
C = math.sin(math.radians(φ_Location)) * math.cos(m
    ath.radians(Difference_Longitude))
D = A/(B-C)
θ = math.degrees(math.atan(D))

#Determine the Azimuth of the Qibla
if Difference_Longitude > 180:
    delta_λ = 360 - Difference_Longitude
else:
    delta_λ = Difference_Longitude

if θ > 0:
   if λ_Location > λ_Kaabah:
       quadrant = "UB" # Utara Barat
   elif λ_Location <= λ_Kaabah:
       quadrant = "UT" # Utara Timur
   elif λ_Location < 0:
       if c >= 180:
           quadrant = "UB"
       else:
           quadrant = "UT"
elif θ < 0:
   if λ_Location > λ_Kaabah:
       quadrant = "SB" # Selatan Barat
```

```
        elif λ_Location <= λ_Kaabah:
            quadrant = "ST" # Selatan Timur
        elif λ_Location < 0:
            if c >= 180:
                quadrant = "SB"
            else:
                quadrant = "ST"

    if quadrant == "UB":
        azimuth_kiblat = 360 - θ
    elif quadrant == "SB":
        azimuth_kiblat = 180 - θ
    elif quadrant == "UT":
        azimuth_kiblat = θ
    elif quadrant == "ST":
        azimuth_kiblat = 180 + θ
    print(azimuth_kiblat)
    139.01065227776655
```

The qibla direction is 139.01. Then, determine the Equation of Time. Install pvlib first.

```
pip install pvlib
```

Then run the code,

```
import pvlib
from datetime import datetime
import math

Year = 2025
Month = 7
Day = 30

# define the date
date = datetime(Year, Month, Day)

# calculate the day of the year
day_of_year = date.timetuple().tm_yday
EoT = pvlib.solarposition.equation_of_time_pvcdrom(
day_of_year) / 60

print(f"Equation of Time on date {date} is {EoT}")
Equation of Time on date 2025-07-30 00:00:00 is
-0.10191142861300423
```

The Equation of Tme is −0.1019, then determine the solar declination. Install skyfield first.

```
pip install skyfield
```

then run the code to calculate Sun Declination

```
#Determine the Sun Declination
#Import Necessary Function
from skyfield.api import load
from skyfield.api import N, S, E, W, wgs84
#Load Ephemeris Data and Planet Objects
ts = load.timescale()
eph = load('de440s.bsp')
planets = load('de440s.bsp')
earth = planets['earth']
sun = planets['sun']

#insert the variable
name_city = "Oslo "
lat_location = 59.913333
long_location = 10.738889
year = 2025
month = 7
day = 30
tz = 1
location = earth + wgs84.latlon(lat_location, long_
location, elevation_m=0)
astro = location.at(ts.utc(year, month, day)).
observe(sun)
app = astro.apparent()
app = astro.apparent()
ra_dec,dec_app,d_app = app.radec()
print(dec_app)
+18deg 34' 49.6"
```

Determine the time of Rashdul Kiblat

```
# Rashdul Qiblat Computation
U = math.degrees(math.atan(1 / (math.tan(math.
    radians(-azimuth_kiblat)) * math.sin(math.
    radians(φ_Location)))))
T_U = math.degrees(math.acos(math.tan(math.radians(
      dec_app.degrees)) * math.cos(math.radians(U)) /
      math.tan(math.radians(lat_location))))
```

```
if U > 0:
   t_HMS = (-abs(T_U) + U) / 15
else:
   t_HMS = (abs(T_U) + U) / 15
R=12+t_HMS
if tz > 0:
    Rashdul_Kiblat = R - EoT + (tz * 15 - long_
    location) / 15
else:
   Rashdul_Kiblat = R - EoT + (abs(tz * 15) -
   abs(long_location)) / 15
# To Convert in Degree Form
degrees = int(Rashdul_Kiblat)
decimal_part = Rashdul_Kiblat - degrees
minutes_total = decimal_part * 60
minutes = int(minutes_total)
# Step 3: Convert the decimal part of minutes to
seconds
seconds = round((minutes_total - minutes) * 60)
print(f"{degrees}° {minutes}' {seconds}"")
10° 22' 17"
```

Rashdul Qiblat occurs at 10:22:17 in Osla on July 30, 2025.

Exercise 4

Use the Python program you've made to calculate the Rashdul Qibla for the city of Washington, DC, in the USA that has geographical latitude of 38.904722 North and geographical longitude of 77.016389 West on March 25, 2025, that has Time zone UTC-5.

Exercise 5

Use the Python program you've made to calculate the Rashdul Qibla for the city of Johannesburg in South Africa that has geographical latitude of 26.204444 South and geographical longitude of 28.045556 East on November 3, 2025, that has Time zone UTC+2.

Prayer Times Calculation 7

Prayer time calculation is based on the position of the sun. Each country has different methods of prayer time calculation. In Malaysia, the prayer time is calculated based on tabular extrapolation on the position of the solar, its hour angle, and Equation of Time.

The tabular extrapolation means that the calculation requires manual computation and cannot be automated for yearly determinations. Therefore, the calculation of prayer times in this does not use tabular extrapolation; instead, it is extracted directly from the sun's position, using the Skyfield library.

ZUHUR

The earliest time for Zuhr begins after solar noon, when the sun passes the local meridian and reaches its highest point in the sky. It is based on Prophet saying:

> *The time for Zuhr is when the sun has passed its zenith until a person's shadow is equal in length to his height.*
>
> Sahih Muslim (612)

The earliest time for Zuhr (Dhuhr) prayer begins just after the sun passes its zenith, known as the solar transit. To ensure the sun has fully crossed the meridian and begun its descent, an offset equal to half of the sun's apparent diameter is applied. This corresponds to approximately 1 minute and 4 seconds.

Using the Skyfield library in Python, the solar daily transit time can be accurately calculated based on the actual position of the sun. This method does not require the Equation of Time, as Skyfield already accounts for the sun's apparent motion and Earth's orbital variations. To calculate the Zuhr prayer

DOI: 10.1201/9781003649120-7

time. First, we need the corresponding location and date. For this case, we use London coordinates, which are 51.5072 North, 0.1276° West, time zone of GMT+1, during the date of 28 May 2025. Then, we need to import function that are needed for the calculation.

```
from skyfield.api import load
from skyfield.api import N, S, E, W, wgs84
from skyfield import almanac
```

The import almanac is added since the function to find solar transit is located under the function almanac.

Put input into the variable

```
latitude = 51.5072
longitude = -0.1276
timezone = 1
day = 28
month = 5
year = 2025
```

Load ephemeris.

```
ts = load.timescale()
eph = load('de440s.bsp')
planets = load('de440s.bsp')
earth = planets['earth']
sun = planets['sun']
```

Feed info of user location

```
location = earth + wgs84.latlon(lat_location, long_
location, elevation_m=0)
```

Skyfield require range of calculate date. If we want to locate the time of the transit between 29 May 2025 to 30 May 2025, this is written as;

```
t0 = ts.utc(year, month, day)
t1 = ts.utc(year, month, day + 1)
```

Then, to find the time of solar transit

```
t = almanac.find_transits(location, sun, t0, t1)
print(t)
     <Time tt=[2460823.99928731]>
```

The result is in Julian Day Number. The extract hour, minute and seconds of solar transit from

```
hour_solar_transit = t.utc.hour
minutes_solar_transit = t.utc.minute
second_solar_transit = t.utc.second

print(hour_solar_transit)
print(minutes_solar_transit )
print(second_solar_transit)
[11.]
[57.]
[49.23954726]
```

This means that solar transit will occur at 11:57:49. However, this is not time-zone corrected. In addition, the Zuhur will occur at 1 minute 4 seconds after solar transit, which is 0.017778 in hour. Therefore, to make the correction based on actual Zuhur time.

```
zuhur_time = hour_solar_transit + (minutes_solar_
transit / 60) + (second_solar_transit / 3600 ) +
timezone + 0.017778
print(zuhur_time)
[12.98145565]
```

This means that after timezone and 1 minute 4 seconds correction, the Zuhr time is [12.98145565]. This is not in time format; to convert into time format:

```
# To Convert in Degree Form
degrees = int(zuhur_time )
decimal_part = float(zuhur_time) - degrees
minutes_total = decimal_part * 60
minutes = int(minutes_total)

seconds = round((minutes_total - minutes) * 60)

sun_astro = location.at(ts.utc(year, month, day,
hour_solar_transit, minutes_solar_transit, second_
solar_transit)).observe(sun)
sun_alt, _, _ = sun_astro.apparent().altaz()

# Check if the sun is above the horizon at Zuhr time
if sun_alt.degrees <= 0:
    zuhur = "Zuhur Does Not Occur"
else:
```

```
    zuhur = f"Zuhur Occurs at {degrees}° {minutes}'
    {seconds}""

print(zuhur)
Zuhur Occurs at 12° 58' 53"
```

This means that the zuhur prayer time at 51.5072 North, 0.1276° West, time zone of GMT+1, during the date of 28 May 2025 occurs at 12:58:53. The above code include counter measure should the Zuhur be calculated at a location near the North or South pole.

Exercise 1

Determine the Zuhr prayer time for Abuja, Nigeria, which is located at coordinates 9.0563° North, 7.4985° East, with a time zone of GMT+1, on the date of 1 January 2014.

Install skyfield

```
pip install skyfield
```

Import Necessary Function

```
from skyfield.api import load
from skyfield.api import N, S, E, W, wgs84
from skyfield import almanac
```

Load Ephemeris Data and Planet Objects

```
ts = load.timescale()
eph = load('de440s.bsp')
planets = load('de440s.bsp')
earth = planets['earth']
sun = planets['sun']
```

Input the Variable

```
lat_location = 9.0563
long_location = 7.4985
timezone = 1
day = 1
month = 1
year = 2014
```

Input info to location variable

```
location = earth + wgs84.latlon(lat_location, long_
location, elevation_m=0)
```

Determine the Range of Data

```
t0 = ts.utc(year, month, day)
t1 = ts.utc(year, month, day + 1)
```

Find the time of Transit

```
t = almanac.find_transits(location, sun, t0, t1)
```

Extract Hour, Minutes, Seconds

```
hour_solar_transit = t.utc.hour
minutes_solar_transit = t.utc.minute
second_solar_transit = t.utc.second

print(hour_solar_transit)
print(minutes_solar_transit )
print(second_solar_transit)
```

Zuhur Time in decimal

```
zuhur_time = hour_solar_transit + (minutes_solar_
transit / 60) + (second_solar_transit / 3600 ) +
timezone + 0.017778
print(zuhur_time)
```

Zuhur in Time Format

```
degrees = int(zuhur_time )
decimal_part = zuhur_time - degrees
minutes_total = decimal_part * 60
minutes = int(minutes_total)

seconds = round((minutes_total - minutes) * 60)

sun_astro = location.at(ts.utc(year, month, day,
hour_solar_transit, minutes_solar_transit, second_
solar_transit)).observe(sun)
sun_alt, _, _ = sun_astro.apparent().altaz()
# Check if the sun is above the horizon at zuhur time
```

```
if sun_alt.degrees <= 0:
     zuhur = "Zuhur Does Not Occur"
else:
    zuhur = f"Zuhur Occurs at {degrees}° {minutes}'
    {seconds}""

print(zuhur)
Zuhur occurs at 12° 34' 37"
```

This means that Zuhur prayer time at Abuja, Kenya coordinate, which are 9.0563 North, 7.4985 East, time zone of GMT+1, during the date of 1 January 2014 will occur at 13:05:08.

Exercise 2

Determine the Zuhr prayer time for Buenos Aires, Argentina, which is located at coordinates 34.6037° South, 58.3816° West, with a time zone of GMT–3, on the date of 5 November 2023.

Exercise 3

Determine the Zuhr prayer time for Vancouver, Canada, which is located at coordinates 49.2827° North, 123.1207° West, with a time zone of GMT–8, on the date of 10 September 2020.

ASAR

The beginning of Asar prayer time is based on the length of the sun's shadow. This is based on the prophetic saying;

> *...Then he prayed Asr when everything was similar (to the length of) its shadow...*
> *...Then he prayed Asr when the shadow of everything was about twice as long as it...*

There are several interpretations for the actual length of shadow for the start of the Asar prayer time. First, as adopted by the Asy-Syafie School of Islamic Thought, the length of the sun shadow for Asar prayer times is

$$\textbf{\textit{Asar Shadow Syafie}} = 1 \times \textbf{\textit{height of a stick}} \quad \text{Equation 7.1}$$

Or written in Python as

```
sun_shadow_asar = 1
```

because we assume the height of the stick is 1, regardless of the unit. For Hanafi School of Islamic Thought, the length of the sun shadow for Asar prayer times is

$$Asar\ Shadow\ Hanafi = 2 \times height\ of\ a\ stick \quad \text{Equation 7.2}$$

Or written in Python as

```
sun_shadow_asar = 2
```

Other opinion, as adopted by Malaysia, the length of the sun shadow for Asar prayer time is

$$Asar\ Shadow\ Others = Sun\ Shadow\ during\ solar\ transit + 1\ height\ of\ a\ stick$$

Equation 7.3

Or written in Python as

```
sun_shadow_asar = 1 + sun_shadow_transit
```

To calculate the prayer time of Asar, we need to calculate the length of the sun shadow in a time loop from solar transit until the stipulated length of the sun shadow for Asar prayer time. For example, for the London coordinate, which are 51.5072 North, 0.1276° West, time zone of GMT+1, during the date of 28 May 2025, where the Asar sun shadow is Equation 7.3. First, we need to import function that are needed for the calculation.

```
from skyfield.api import load
from skyfield.api import N, S, E, W, wgs84
from skyfield import almanac
import math
```

Put input into the variable

```
lat_location = 51.5072
long_location = -0.1276
```

```
timezone = 1
day = 28
month = 5
year = 2025
```

Load ephemeris.

```
ts = load.timescale()
eph = load('de440s.bsp')
planets = load('de440s.bsp')
earth = planets['earth']
sun = planets['sun']
```

Feed info of user location

```
location = earth + wgs84.latlon(lat_location, long_
location, elevation_m=0)
```

Skyfield require range of calculate date. If we want to locate the time of the transit between 29 May 2025 to 30 May 2025, this is written as;

```
t0 = ts.utc(year, month, day)
t1 = ts.utc(year, month, day + 1)
```

Then, to find the time of solar transit

```
t = almanac.find_transits(location, sun, t0, t1)
```

After finding the solar transit, determine the position of sun altitude at the time of the solar transit

```
h, m, s = t.utc.hour, t.utc.minute, t.utc.second
sun_astro = location.at(ts.utc(year, month, day,
h, m)).observe(sun)
sun_app = sun_astro.apparent()
sun_alt, sun_az, distance = sun_app.altaz()
```

In the above code, sun_astro is the astrometric position of the sun. The astrometric position needs to be converted to apparent position to determine the altitude of the sun. Therefore, the code sun_app = sun_astro.apparent() makes the conversion from astrometric to apparent position. Then, to determine the altitude of the sun, sun_alt, sun_az, distance = sun_app.altaz(), where sun_alt is the sun altitude, sun_az is the sun azimuth. distance is the apparent distance between the sun and earth.

```
print(sun_alt)
60deg 02′ 14.0″
```

From the above code, we know that the altitude of the sun is 60° 02' 14.0". Then, determine the length of the sun shadow during transit, which is written as

```
sun_shadow_transit = 1/ (math.tan(math.radians(sun_alt
.degrees)))
print(sun_shadow_transit)
0.576484553268423
```

The result is 0.5764, indicating that at the time of solar transit, the length of the sun's shadow is approximately 57.64% of the stick's height, assuming the stick height is 1. Then to determine the sun shadow has reach the length from Equation 7.3.

```
sun_shadow_asar = 1 + sun_shadow_transit
print(sun_shadow_asar)
1.5764845532684229
```

At the time of solar transit, the length of the sun's shadow is approximately 0.5764, meaning it is only 57.64% of the stick's height. However, for the Asar prayer, the required shadow length is 1.5764, or 157.64% of the stick's height. This indicates that the sun must descend further in the sky before the Asar time begins. To accurately determine the time for Asar prayer, a time-based loop is implemented starting from the moment of solar transit. This loop checks the length of the shadow at each point in time until it reaches the required Asar length. To make the process more efficient, the loop is divided into three stages: hour loop, minute loop, and second loop, allowing the program to incrementally and precisely identify the exact time when the condition is met. To develop with hour time loop, with the rule of the loop the sun shadow does not pass the length of the Asar sun shadow,

```
# Start with hour
test = 1
while True:

  # Calculate the shadow length
  sun_astro = location.at(ts.utc(year, month, day, h,
  m, s)).observe(sun)
  sun_alt, _, _ = sun_astro.apparent().altaz() # Get
  the altitude of the sun
  sun_shadow= 1 / math.tan(math.radians(sun_alt.
  degrees))
```

```
  if sun_alt.degrees <= 0:
    break
  if test > 24:
    break
  if sun_shadow >= sun_shadow_asar:
    break # Exit the loop if the shadow length matches
    or exceeds the desired length
  h += 1
  test +=1
  print(f'Sun Altitude {sun_alt.degrees}, Sun_Shadow
  {sun_shadow} at hour {h}')
```

```
Sun Altitude [58.06080941], Sun_Shadow
0.6233945983146184 at hour [12.]
Sun Altitude [60.03162483], Sun_Shadow
0.5766145603035456 at hour [13.]
Sun Altitude [57.62354596], Sun_Shadow
0.6340429848026038 at hour [14.]
Sun Altitude [51.70606881], Sun_Shadow
0.7895804424162487 at hour [15.]
Sun Altitude [43.7299877], Sun_Shadow
1.0453443161558726 at hour [16.]
Sun Altitude [34.76344347], Sun_Shadow
1.4407721100751296 at hour [17.]
```

From the printed output above, we can see that the Sun's shadow exceeds the required Asar shadow length at hour 17. Therefore, we step back one hour and proceed with a minute-level time loop to more precisely determine the exact minute when the required shadow length is first reached.

```
# Once the condition is met for hours, move to minutes
h_asar = h - 1
test=1
while True:
  # Calculate the shadow length
  sun_astro = location.at(ts.utc(year, month, day,
  h_asar, m, s)).observe(sun)
  sun_alt, _, _ = sun_astro.apparent().altaz() # Get
  the altitude of the sun
  sun_shadow = 1 / math.tan(math.radians(sun_alt.
  degrees))

  if sun_alt.degrees <= 0:
    break
  if test > 1440:
    break
```

```
    if sun_shadow >= sun_shadow_asar:
      break # Exit the loop if the shadow length matches
      or exceeds the desired length
    m += 1
    test +=1
    print(f'Sun Altitude {sun_alt.degrees}, Sun_Shadow
    {sun_shadow} at minute {m}')

    # Increment time in minutes
  m_asar = m - 1

  Sun Altitude [34.76344347], Sun_Shadow
  1.4407721100751296 at minute 2
  Sun Altitude [34.60973771], Sun_Shadow
  1.449055572925344 at minute 3
  Sun Altitude [34.45594556], Sun_Shadow
  1.4574083986033666 at minute 4
  Sun Altitude [34.30206933], Sun_Shadow
  1.4658314538010262 at minute 5
  Sun Altitude [34.14811128], Sun_Shadow
  1.4743256209720872 at minute 6
  Sun Altitude [33.99407369], Sun_Shadow
  1.4828917986639925 at minute 7
  Sun Altitude [33.8399588], Sun_Shadow
  1.4915309018584026 at minute 8
  Sun Altitude [33.68576885], Sun_Shadow
  1.50024386232121 at minute 9
  Sun Altitude [33.53150605], Sun_Shadow
  1.5090316289614694 at minute 10
  Sun Altitude [33.37717262], Sun_Shadow
  1.517895168199992 at minute 11
  Sun Altitude [33.22277074], Sun_Shadow
  1.5268354643482442 at minute 12
  Sun Altitude [33.06830259], Sun_Shadow
  1.5358535199970826 at minute 13
  Sun Altitude [32.91377034], Sun_Shadow
  1.544950356416044 at minute 14
  Sun Altitude [32.75917614], Sun_Shadow
  1.554127013963914 at minute 15
  Sun Altitude [32.60452212], Sun_Shadow
  1.5633845525101653 at minute 16
  Sun Altitude [32.44981041], Sun_Shadow
  1.5727240518677097 at minute 17
```

From the printed output above, we can see that the Sun's shadow exceeds the required Asar shadow length at minute 23. Therefore, we step back one minute and proceed with a second-level time loop to more precisely determine the exact minute when the required shadow length is first reached.

```
# Once the condition is met for minutes, move to
seconds
test=1
while True:
  # Calculate the shadow length
  sun_astro = location.at(ts.utc(year, month, day,
  h_asar, m_asar, s)).observe(sun)
  sun_alt, _, _ = sun_astro.apparent().altaz() # Get
  the altitude of the sun
  sun_shadow = 1 / math.tan(math.radians(sun_alt.
  degrees))

  if sun_alt.degrees <= 0:
    break
  if test > 86400:
    break
  if sun_shadow >= sun_shadow_asar:
    break # Exit the loop if the shadow length matches
    or exceeds the desired length
  s += 1
  test +=1
  print(f'Sun Altitude {sun_alt.degrees}, Sun_Shadow
  {sun_shadow} at second {s}')

  # Increment time in seconds

s_asar = s

Sun Altitude [32.44981041], Sun_Shadow
1.5727240518677097 at second 2
Sun Altitude [32.44723141], Sun_Shadow
1.5728804108174916 at second 3
Sun Altitude [32.44465239], Sun_Shadow
1.573036792844485 at second 4
Sun Altitude [32.44207335], Sun_Shadow
1.5731931979536897 at second 5
Sun Altitude [32.4394943], Sun_Shadow
1.573349626150443 at second 6
Sun Altitude [32.43691523], Sun_Shadow
1.5735060774396996 at second 7
Sun Altitude [32.43433615], Sun_Shadow
1.573662551826798 at second 8
```

```
Sun Altitude [32.43175705], Sun_Shadow
1.5738190493167965 at second 9
Sun Altitude [32.42917794], Sun_Shadow
1.573975569914748 at second 10
Sun Altitude [32.42659881], Sun_Shadow
1.5741321136259998 at second 11
Sun Altitude [32.42401967], Sun_Shadow
1.5742886804555098 at second 12
Sun Altitude [32.42144051], Sun_Shadow
1.5744452704086311 at second 13
Sun Altitude [32.41886133], Sun_Shadow
1.5746018834903772 at second 14
Sun Altitude [32.41628214], Sun_Shadow
1.5747585197059564 at second 15
Sun Altitude [32.41370294], Sun_Shadow
1.5749151790605282 at second 16
Sun Altitude [32.41112372], Sun_Shadow
1.5750718615593058 at second 17
Sun Altitude [32.40854448], Sun_Shadow
1.5752285672074033 at second 18
Sun Altitude [32.40596523], Sun_Shadow
1.5753852960098962 at second 19
Sun Altitude [32.40338597], Sun_Shadow
1.5755420479721345 at second 20
Sun Altitude [32.40080668], Sun_Shadow
1.5756988230991047 at second 21
Sun Altitude [32.39822739], Sun_Shadow
1.5758556213961656 at second 22
Sun Altitude [32.39564808], Sun_Shadow
1.5760124428683018 at second 23
Sun Altitude [32.39306875], Sun_Shadow
1.5761692875208748 at second 24
Sun Altitude [32.39048941], Sun_Shadow
1.5763261553589674 at second 25
Sun Altitude [32.38791005], Sun_Shadow
1.5764830463876598 at second 26
```

From the printed output above, we can see that the sun's shadow exceeds the required Asar shadow length at seconds 26. Therefore, we step back one second. The Asar time, after the time loop, with timezone correction is

```
asar_time = (h_asar + (m_asar) / 60 + s_asar / 3600) +
timezone
print(asar_time)
[17.27388889]
```

This is observed in decimal. To convert to time format.

```
asar_time = float(asar_time)
degrees = int(asar_time)
decimal_part = asar_time - degrees
minutes_total = decimal_part * 60
minutes = int(minutes_total)

seconds = round((minutes_total - minutes) * 60)

if sun_alt.degrees <= 0 or test >86400:
    asar = "Asar Does Not Occur"
else:
    asar = f"Asar Occurs at {degrees}° {minutes}'
    {seconds}""

print(asar)
Asar Occurs at 17° 16' 26"
```

Therefore, asar prayer time at London, during 28 May 2025, based on Equation 7.3 occurs at 17:16:26. The second part is a counter measure should the calculated position is near the North or South Pole. Now, use the same London coordinate, which are 51.5072 North, 0.1276° West, time zone of GMT+1, during the date of 28 May 2025. Determine the prayer time of asar, where the asar sun shadow is Equation 7.1. First, we need to import function that are needed for the calculation.

```
from skyfield.api import load
from skyfield.api import N, S, E, W, wgs84
from skyfield import almanac
import math
```

Put input into the variable

```
lat_location = 51.5072
long_location = -0.1276
timezone = 1
day = 28
month = 5
year = 2025
```

Load ephemeris.

```
ts = load.timescale()
eph = load('de440s.bsp')
```

```
planets = load('de440s.bsp')
earth = planets['earth']
sun = planets['sun']
```

Feed info of user location

```
location = earth + wgs84.latlon(lat_location, long_
location, elevation_m=0)
```

Skyfield require range of calculate date, this is written as;

```
t0 = ts.utc(year, month, day)
t1 = ts.utc(year, month, day + 1)
```

Then, to find the time of solar transit

```
t = almanac.find_transits(location, sun, t0, t1)
```

After finding the solar transit, determine the position of sun altitude at the time of the solar transit

```
h, m, s = t.utc.hour, t.utc.minute, t.utc.second
sun_astro = observer.at(ts.utc(year, month, day, h,
m)).observe(sun)
sun_app = sun_astro.apparent()
sun_alt, sun_az, distance = sun_app.altaz()
print(sun_alt)
60deg 02' 14.0"
```

Then, determine the length of the sun shadow during transit, which is written as

```
sun_shadow_transit = 1/ (math.tan(math.radians(sun_alt
.degrees)))
print(sun_shadow_transit)
0.576484553268423
```

Then to determine the sun shadow has reach the length from Equation 7.1.

```
sun_shadow_asar = 1
```

At the time of solar transit, the length of the sun's shadow is approximately 0.57, meaning it is only 57% of the stick's height. However, for the Asar prayer, the required shadow length is 1 or 100% of the stick's height. This indicates that the sun must descend further in the sky before the Asar time begins. Write the hour time loop, with the rule of the loop the sun shadow does not pass the length of the afar sun shadow

```
# Start with hour
test=1
while True:

  # Calculate the shadow length
  sun_astro = location.at(ts.utc(year, month, day, h,
  m, s)).observe(sun)
  sun_alt, _, _ = sun_astro.apparent().altaz() # Get
  the altitude of the sun
  sun_shadow= 1 / math.tan(math.radians(sun_alt.
  degrees))

  if sun_alt.degrees <= 0:
    break
  if test > 24:
    break

  if sun_shadow >= sun_shadow_asar:
    break # Exit the loop if the shadow length matches
    or exceeds the desired length
  h += 1
  print(f'Sun Altitude {sun_alt.degrees}, Sun_Shadow
  {sun_shadow} at hour {h}')
```

Proceed with a minute-level time loop to more precisely determine the exact minute when the required shadow length is first reached.

```
# Once the condition is met for hours, move to minutes
h_asar = h - 1
test=1

while True:
  # Calculate the shadow length
  sun_astro = location.at(ts.utc(year, month, day,
  h_asar, m, s)).observe(sun)
  sun_alt, _, _ = sun_astro.apparent().altaz() # Get
  the altitude of the sun
  sun_shadow = 1 / math.tan(math.radians(sun_alt.
  degrees))
```

```
    if sun_alt.degrees <= 0:
      break
    if test > 1440:
      break

    if sun_shadow >= sun_shadow_asar:
      break # Exit the loop if the shadow length matches
      or exceeds the desired length
    m += 1
    print(f'Sun Altitude {sun_alt.degrees}, Sun_Shadow
    {sun_shadow} at minute {m}')
    # Increment time in minutes
  m_asar = m - 1
```

Proceed with a second-level time loop to more precisely determine the exact minute when the required shadow length is first reached.

```
  # Once the condition is met for minutes, move to
  seconds
  test = 1
  while True:
    # Calculate the shadow length
    sun_astro = location.at(ts.utc(year, month, day,
    h_asar, m_asar, s)).observe(sun)
    sun_alt, _, _ = sun_astro.apparent().altaz() # Get
    the altitude of the sun
    sun_shadow = 1 / math.tan(math.radians(sun_alt.
    degrees))
    if sun_alt.degrees <= 0:
      break
    if test > 86400:
      break

    if sun_shadow >= sun_shadow_asar:
    break # Exit the loop if the shadow length matches
    or exceeds the desired length
    s += 1
    print(f'Sun Altitude {sun_alt.degrees}, Sun_Shadow
    {sun_shadow} at second {s}')

    # Increment time in seconds

  s_asar = s
```

The asar time, after the time loop, with timezone correction is

```
asar_time = (h_asar + (m_asar) / 60 + s_asar / 3600) +
timezone
print(asar_time)
```

This is observed in decimal. To convert to time format.

```
asar_time = float(asar_time)
degrees = int(asar_time)
decimal_part = asar_time - degrees
minutes_total = decimal_part * 60
minutes = int(minutes_total)

seconds = round((minutes_total - minutes) * 60)
if sun_alt.degrees <= 0 or test >86400:
    asar = "Asar Does Not Occur"
else:
    asar = f"Asar Occurs at {degrees}° {minutes}'
    {seconds}""

print(asar)
Asar Occurs at 15° 52' 6"
```

Therefore, asar prayer time at London, during 28 May 2025, based on Equation 7.1 occurs at 15:52:06. Finally, use the same London coordinate, which are 51.5072 North, 0.1276° West, time zone of GMT+1, during the date of 28 May 2025. Determine the prayer time of asar, where the asar sun shadow is Equation 7.2. Following from the above code, the change is Equation 7.2. Where,

```
sun_shadow_asar = 2
```

The hour time loop, with the rule of the loop the sun shadow does not pass the length of the afar sun shadow

```
# Start with hour
test = 1
while True:

  # Calculate the shadow length
  sun_astro = location.at(ts.utc(year, month, day, h,
  m, s)).observe(sun)
  sun_alt, _, _ = sun_astro.apparent().altaz() # Get
  the altitude of the sun
```

```
    sun_shadow= 1 / math.tan(math.radians(sun_alt.
    degrees))

    if sun_alt.degrees <= 0:
      break
    if test > 24:
      break

    if sun_shadow >= sun_shadow_asar:
      break # Exit the loop if the shadow length matches
      or exceeds the desired length
    h += 1

  # Once the condition is met for hours, move to minutes
  h_asar = h - 1
  test=1
  while True:
    # Calculate the shadow length
    sun_astro = location.at(ts.utc(year, month, day,
    h_asar, m, s)).observe(sun)
    sun_alt, _, _ = sun_astro.apparent().altaz() # Get
    the altitude of the sun
    sun_shadow = 1 / math.tan(math.radians(sun_alt.
    degrees))

    if sun_alt.degrees <= 0:
      break
    if test > 1440:
      break

    if sun_shadow >= sun_shadow_asar:
      break # Exit the loop if the shadow length matches
      or exceeds the desired length
    m += 1

    # Increment time in minutes
  m_asar = m - 1
  test = 1

  # Once the condition is met for minutes, move to
  seconds
  while True:
    # Calculate the shadow length
    sun_astro = location.at(ts.utc(year, month, day,
    h_asar, m_asar, s)).observe(sun)
```

```
        sun_alt, _, _ = sun_astro.apparent().altaz() # Get
        the altitude of the sun
        sun_shadow = 1 / math.tan(math.radians(sun_alt.
        degrees))

        if sun_alt.degrees <= 0:
          break
        if test > 86400:
          break

        if sun_shadow >= sun_shadow_asar:
          break # Exit the loop if the shadow length matches
          or exceeds the desired length
        s += 1

        # Increment time in seconds

    s_asar = s
```

The asar time, after the time loop, with timezone correction is

```
asar_time = (h_asar + (m_asar) / 60 + s_asar / 3600) +
timezone
print(asar_time)
```

This is observed in decimal. To convert to time format.

```
asar_time = float(asar_time)
degrees = int(asar_time)
decimal_part = asar_time - degrees
minutes_total = decimal_part * 60
minutes = int(minutes_total)
seconds = round((minutes_total - minutes) * 60)
print(f"{degrees}° {minutes}' {seconds}"")
17° 53' 55"
```

Therefore, asar prayer time at London, during 28 May 2025, based on Equation 7.2 occurs at 17:53:55.

Exercise 1

Using the coordinates of Cairo, which are 30.0444° North, 31.2357° East, and a time zone of GMT+2, determine the Asar prayer time for the date 10 March 2025.

Use Equation 14 to compute the shadow length required at Asar.

Exercise 2

Using the coordinates of Kuala Lumpur, which are 3.1390° North, 101.6869° East, and a time zone of GMT+8, determine the Asar prayer time for the date 21 September 2025.

Apply Equation 15 to evaluate the required shadow length for Asar.

Exercise 3

Using the coordinates of New York City, which are 40.7128° North, 74.0060° West, and a time zone of GMT–4, determine the Asar prayer time for the date August 5, 2025.

Use Equation 16 to find when the shadow length meets the Asar requirement.

MAGHRIB

The beginning of Maghrib prayer time is based on the position of the sunset. This is based on the prophetic saying;

> *...Then he prayed Maghrib when the sun had set and the fasting person breaks fast....*

The position of the sun must be completely below the horizon to mark the beginning of Maghrib prayer. This means that the upper limb of the sun must have fully disappeared. However, factors such as atmospheric refraction and the observer's elevation above sea level can affect the apparent position of the sun at sunset. These variables must be carefully considered when calculating the precise time for Maghrib prayer. The determination of Maghrib prayer time is based on the find_settings function. Skyfield uses the official definition of sunrise and sunset from the United States Naval Observatory, which defines them as the moment when the center of the sun is 50 arcminutes below the horizon, to account for both the average solar radius of 16 arcminutes and for roughly 34 arcminutes of atmospheric refraction at the horizon. To determine the time of the maghrib prayer time, assuming the user is located at Sydney Tower Eye, with Latitude 33.8688° South, Longitude 151.2093° East, on 23

March 2025, where the timezone in effect is GMT + 11, with elevation of 350 meter. First, we need to import function that are needed for the calculation.

```
from skyfield.api import load
from skyfield.api import N, S, E, W, wgs84
from skyfield import almanac
import math
```

Put input into the variable

```
lat_location = -33.8688
long_location = 151.2093
timezone = 11
day = 23
month = 3
year = 2025
ele = 350
```

Load ephemeris.

```
ts = load.timescale()
eph = load('de440s.bsp')
planets = load('de440s.bsp')
earth = planets['earth']
sun = planets['sun']
```

Feed info of user location

```
location = earth + wgs84.latlon(lat_location, long_
location, elevation_m=ele)
```

Skyfield require range of calculate date, this is written as;

```
t0 = ts.utc(year, month, day)
t1 = ts.utc(year, month, day + 1)
```

Then, include the variable of refraction and location elevation

```
from skyfield.units import Angle
from numpy import arccos
from skyfield.earthlib import refraction

altitude_m = ele
earth_radius_m = 6378136.6
```

```
side_over_hypotenuse = earth_radius_m / (earth_
radius_m + altitude_m)
h = Angle(radians=-arccos(side_over_hypotenuse))
solar_radius_degrees = 16 / 60
r = refraction(0.0, temperature_C=15.0,
pressure_mbar=1030.0)
```

Then, determine the time of the sunset

```
t, y = almanac.find_settings(location, sun, t0, t1,
horizon_degrees=-r + h.degrees - solar_radius_degrees)
h, m, s = t.utc.hour, t.utc.minute, t.utc.second
print(h,m,s)
[8.] [5.] [28.44139129]
```

The time of the sunset is 8:05:28.44, at UTC time. To convert into local time with GMT + 11.

```
maghrib_time = float(h + m / 60 + s / 3600 + timezone)
print(maghrib_time)
19.09123372
```

When calculating prayer times near sunset, there may be instances where the computed time exceeds 24 hours. This typically occurs due to the addition of the local time zone offset to the base time (often in UTC). To correct this and ensure the resulting prayer time remains within a 24-hour format, the following code is used to normalize the time

```
maghrib_time %= 24 # Ensure 24-hour clock format
print(maghrib_time)
```

From this moment, we can convert into time format,

```
maghrib_time = float(maghrib_time)
degrees = int(maghrib_time)
decimal_part = maghrib_time - degrees
minutes_total = decimal_part * 60
minutes = int(minutes_total)
seconds = round((minutes_total - minutes) * 60)

sun_astro = location.at(ts.utc(year, month, day, h, m,
s)).observe(sun)
sun_alt, _, _ = sun_astro.apparent().altaz()
if sun_alt.degrees >= 0:
    maghrib = "Maghrib Does Not Occur"
```

```
else:
    maghrib = f"Maghrib Occurs at {degrees}° {minutes}'
    {seconds}""
print(maghrib)
Maghrib Occurs at 19° 5' 28"
```

This means that Maghrib at Sydney Tower Eye, with Latitude 33.8688° South, Longitude 151.2093° East, on 23 March 2025, where the timezone in effect is GMT + 11, with elevation of 350 meter occur at 19:5:28.

Exercise 1

Determine the Maghrib prayer time on 15 July 2025 for a user located at the KL Tower, Malaysia, with the following details:

- Latitude: 3.1579° North
- Longitude: 101.7123° East
- Elevation: 300 meters
- Time Zone: GMT +8

Exercise 2

Calculate the Maghrib prayer time for Cairo, Egypt on 10 October 2025, with the following coordinates:

- Latitude: 30.0444° North
- Longitude: 31.2357° East
- Elevation: 75 meters
- Time Zone: GMT +2

Exercise 3

On 1 January 2025, determine the time of Maghrib prayer for a user located at the Empire State Building, New York, USA:

- Latitude: 40.7484° North
- Longitude: 73.9857° West
- Elevation: 381 meters
- Time Zone: GMT –5 (Standard Time, no DST)

Exercise 4

Compute the Maghrib prayer time for a location in Makkah, Saudi Arabia, on 28 May 2025:

- Latitude: 21.4225° North
- Longitude: 39.8262° East
- Elevation: 277 meters
- Time Zone: GMT +3

ISYA'

The beginning of Isya' prayer time is based on the position of the sky condition after sunset. This is based on the prophetic saying;

> *...Then he prayed Isha when the twilight had vanished....*

The disappearance of twilight from the sky is caused by the sun's continued descent below the horizon after sunset. At a certain angle, known as the solar depression angle, the sun's rays are no longer refracted by the atmosphere in a way that illuminates the night sky. When this critical angle is reached, the sky becomes completely dark, marking the beginning of the 'Isha (Isya') prayer time. However, Islamic scholars differ in their opinions regarding the exact degree of solar depression that signifies the start of 'Isha. While many adopt angles between 15° and 18° below the horizon, there is no unanimous agreement, and various regions apply different standards based on jurisprudential reasoning and observational studies. To determine the Isya' prayer time, at Bangkok where Latitude 13.7563° North, Longitude 100.5018° East and Time Zone GMT+7, with elevation of 150 meter on 16 June 2025 for 18 degrees of solar depression during Isya'. First, we need to import function that are needed for the calculation.

```
from skyfield.api import load
from skyfield.api import N, S, E, W, wgs84
from skyfield import almanac
import math
```

Put input into the variable

```
lat_location = 13.7563
long_location = 100.5018
```

```
timezone = 7
day = 16
month = 6
year = 2025
ele = 150
```

Load ephemeris.

```
ts = load.timescale()
eph = load('de440s.bsp')
planets = load('de440s.bsp')
earth = planets['earth']
sun = planets['sun']
```

Feed info of user location

```
location = earth + wgs84.latlon(lat_location, long_
location, elevation_m=ele)
```

Skyfield require range of calculate date, this is written as;

```
t0 = ts.utc(year, month, day)
t1 = ts.utc(year, month, day + 1)
```

Then, include the variable of refraction and location elevation

```
from skyfield.units import Angle
from numpy import arccos
from skyfield.earthlib import refraction

altitude_m = ele
earth_radius_m = 6378136.6
side_over_hypotenuse = earth_radius_m / (earth_
radius_m + altitude_m)
h = Angle(radians=-arccos(side_over_hypotenuse))
solar_radius_degrees = 16 / 60
r = refraction(0.0, temperature_C=15.0,
pressure_mbar=1030.0)
```

Then, determine the time of the sunset

```
t, y = almanac.find_settings(location, sun, t0, t1,
horizon_degrees=-r + h.degrees - solar_radius_degrees)
h, m, s = t.utc.hour, t.utc.minute, t.utc.second
print(h,m,s)
[11.] [48.] [36.56323494]
```

Since the 'Isha (Isya') prayer occurs after sunset, it is necessary to perform a time-based loop starting from the moment of sunset and continuing until the solar altitude reaches 18 degrees of solar depression. The time loop operation is similar with the previous asar prayer time loop operation. First determine the solar altitude,

```
sun_astro = location.at(ts.utc(year, month, day, h,
m)).observe(sun)
sun_app = sun_astro.apparent()
sun_alt, sun_az, distance = sun_app.altaz()
print(sun_alt)
-01deg 06' 01.2"
```

To develop with hour time loop, with the rule of the loop the solar degree does not pass the altitude of – 18.

```
# Start with hour
test=1
while True:

  # Calculate the Solar Altitude
  sun_astro = location.at(ts.utc(year, month, day, h,
  m, s)).observe(sun)
  sun_alt, _, _ = sun_astro.apparent().altaz() # Get
  the altitude of the sun
  isha_angle = 18
  elevation_correction = 0.0293 * math.sqrt(ele)
  isha_angle_actual = -isha_angle - elevation_correction

  if sun_alt.degrees >= 0:
    break
  if test > 24:
    break

  if sun_alt.degrees <= isha_angle_actual:
    break # Exit the loop if the solar altitude
    located below -18 degree
  print(f'Sun Altitude {sun_alt.degrees}, at hour
  {h}')
  h += 1
  test += 1
Sun Altitude [-1.23504713], at hour [11.]
Sun Altitude [-14.26400679], at hour [12.]
```

In the above code, **if sun_alt.degrees <= isha_angle_actual:** is used to determine whether the solar altitude has reach below 18 degree of depression. In this example, the elevation factor is considered for high-altitude locations. When calculating prayer times, particularly for Isha and Fajr, the position of the sun below the horizon is critical. Typically, the angle used for Isha is when the sun is 18 degrees below the horizon. However, this standard angle is based on observations made at sea level.

At higher elevations, the situation changes due to the geometry of the Earth and the observer's horizon. When a person is located on a mountain or at a high-altitude location, their view of the sky is broader, and they are physically above a portion of the atmosphere. As a result, the sun appears to set faster, and darkness arrives sooner compared to someone at sea level. To accurately reflect this earlier onset of night, the Isha angle must be adjusted slightly downward, meaning a larger depression angle. This is why we include an *elevation correction factor* in the calculation. The correction is often calculated using the formula:

$$\textbf{elevation correction} = 0.0293 \times \sqrt{(\textbf{elevation in meters})} \qquad \text{Equation 7.4}$$

This value is subtracted from the standard Isha angle (e.g., –18°), giving a slightly steeper actual angle for high-altitude locations. For instance, at 4000 meters elevation, the correction might be around 1.85 degrees, so the adjusted angle becomes approximately –19.85 degrees. This adjustment ensures that the calculated Isha time matches the true observable darkness experienced by someone at that elevation. Without applying this correction, the prayer time would be inaccurately delayed, especially in mountainous regions. From the output, the sun altitude reach angle below –18 degrees at hour 12. Then proceed with a minute-level time loop to more precisely determine the exact minute when the solar altitude is first reached.

```
# Once the condition is met for hours, move to minutes
h_isya = h - 1
test=1

while True:
  # Calculate the Solar Altitude
  sun_astro = location.at(ts.utc(year, month, day,
  h_isya, m, s)).observe(sun)
  sun_alt, _, _ = sun_astro.apparent().altaz() # Get
  the altitude of the sun

  if sun_alt.degrees >= 0:
    break
  if test > 1440:
    break
```

```
    if sun_alt.degrees <= isha_angle_actual:
      break # Exit the loop if the solar altitude
      located below -18 degree
    m += 1
    print(f'Sun Altitude {sun_alt.degrees}, at minute
    {m}')

    # Increment time in minutes
  m_isya = m - 1
  Sun Altitude [-14.26400679], at minute [49.]
  Sun Altitude [-14.47642833], at minute [50.]
  Sun Altitude [-14.68866518], at minute [51.]
  Sun Altitude [-14.90071568], at minute [52.]
  Sun Altitude [-15.11257813], at minute [53.]
  Sun Altitude [-15.32425085], at minute [54.]
  Sun Altitude [-15.53573212], at minute [55.]
  Sun Altitude [-15.74702021], at minute [56.]
  Sun Altitude [-15.95811338], at minute [57.]
  Sun Altitude [-16.16900984], at minute [58.]
  Sun Altitude [-16.37970783], at minute [59.]
  Sun Altitude [-16.59020554], at minute [60.]
  Sun Altitude [-16.80050114], at minute [61.]
  Sun Altitude [-17.0105928], at minute [62.]
  Sun Altitude [-17.22047865], at minute [63.]
  Sun Altitude [-17.43015683], at minute [64.]
  Sun Altitude [-17.63962543], at minute [65.]
  Sun Altitude [-17.84888254], at minute [66.]
  Sun Altitude [-18.05792617], at minute [67.]
  Sun Altitude [-18.26675447], at minute [68.]
```

From the printed output above, we can see that the sun altitude reaches below the required Isya' depression degree at minute 68. Therefore, we step back one minute and proceed with a second-level time loop to more precisely determine the exact minute when the required Isya' depression degree is first reached.

```
  # Once the condition is met for minutes, move to
  seconds
  test = 1
  while True:
    # Calculate the Solar Altitude
    sun_astro = location.at(ts.utc(year, month, day,
    h_isya, m_isya, s)).observe(sun)
    sun_alt, _, _ = sun_astro.apparent().altaz() # Get
    the altitude of the sun
```

```
    if sun_alt.degrees >= 0:
      break
    if test > 86400:
      break

    if sun_alt.degrees <= isha_angle_actual:
      break # Exit the loop if the solar altitude
      located below -18 degree
    s += 1
    print(f'Sun Altitude {sun_alt.degrees}, at second
    {s}')

    # Increment time in seconds

  s_isya = s
```

```
Sun Altitude [-18.26675447], at second [37.56323494]
Sun Altitude [-18.2702331], at second [38.56323494]
Sun Altitude [-18.27371168], at second [39.56323494]
Sun Altitude [-18.27719019], at second [40.56323494]
Sun Altitude [-18.28066865], at second [41.56323494]
Sun Altitude [-18.28414704], at second [42.56323494]
Sun Altitude [-18.28762538], at second [43.56323494]
Sun Altitude [-18.29110365], at second [44.56323494]
Sun Altitude [-18.29458186], at second [45.56323494]
Sun Altitude [-18.29806001], at second [46.56323494]
Sun Altitude [-18.30153811], at second [47.56323494]
Sun Altitude [-18.30501614], at second [48.56323494]
Sun Altitude [-18.30849411], at second [49.56323494]
Sun Altitude [-18.31197202], at second [50.56323494]
Sun Altitude [-18.31544987], at second [51.56323494]
Sun Altitude [-18.31892766], at second [52.56323494]
Sun Altitude [-18.32240539], at second [53.56323494]
Sun Altitude [-18.32588306], at second [54.56323494]
Sun Altitude [-18.32936066], at second [55.56323494]
Sun Altitude [-18.33283821], at second [56.56323494]
Sun Altitude [-18.3363157], at second [57.56323494]
Sun Altitude [-18.33979312], at second [58.56323494]
Sun Altitude [-18.34327049], at second [59.56323494]
Sun Altitude [-18.34674779], at second [60.56323494]
Sun Altitude [-18.35022504], at second [61.56323494]
Sun Altitude [-18.35370222], at second [62.56323494]
Sun Altitude [-18.35717935], at second [63.56323494]
```

From the printed output above, we can see that the solar altitude exceeds the required Isya' Solar Depression at second 63. Therefore, we step back one second. The Isya' time, after the time loop, with timezone correction is

```
isya_time = (h_isya + (m_isya) / 60 + s_isya /
3600)+timezone
print(isya_time)
[20.13444444]
```

To correct this and ensure the resulting prayer time remains within a 24-hour format,

```
isya_time %= 24 # Ensure 24-hour clock format
print(isya_time)
```

To convert into time format,

```
isya_time = float(isya_time)
degrees = int(isya_time)
decimal_part = isya_time - degrees
minutes_total = decimal_part * 60
minutes = int(minutes_total)
seconds = round((minutes_total - minutes) * 60)

if sun_alt.degrees >= 0 or test >86400:
    isya = "Isya' Does Not Occur"
else:
    isya = f"Isya' Occurs at {degrees}° {minutes}'
    {seconds}""

print(isya)
Isya' Occurs at 20° 8' 4"
```

Isya prayer time at Bangkok where Latitude 13.7563° North, Longitude 100.5018° East and Time Zone GMT+7, with elevation of 150 meter on 16 June 2025 for 18 degree of solar depression is Isya' is at 20:08:04.

Exercise 1

Determine the 'Isya prayer time at Jakarta, Indonesia, where Latitude: 6.2088° South, Longitude: 106.8456° East, Time Zone: GMT+7, Elevation: 50 meters, Date: 21 July 2025, Assume 'Isya begins at a solar depression of −17°.

First, we need to import function that are needed for the calculation.

```
from skyfield.api import load
from skyfield.api import N, S, E, W, wgs84
from skyfield import almanac
import math
```

Put input into the variable

```
lat_location = -6.2088
long_location = 106.8456
timezone = 7
day = 21
month = 7
year = 2025
ele = 50
```

Load ephemeris.

```
ts = load.timescale()
eph = load('de440s.bsp')
planets = load('de440s.bsp')
earth = planets['earth']
sun = planets['sun']
```

Feed info of user location

```
location = earth + wgs84.latlon(lat_location, long_
location, elevation_m=ele)
```

Skyfield require range of calculate date, this is written as;

```
t0 = ts.utc(year, month, day)
t1 = ts.utc(year, month, day + 1)
```

Then, include the variable of refraction and location elevation

```
from skyfield.units import Angle
from numpy import arccos
from skyfield.earthlib import refraction

altitude_m = ele
earth_radius_m = 6378136.6
side_over_hypotenuse = earth_radius_m / (earth_
radius_m + altitude_m)
h = Angle(radians=-arccos(side_over_hypotenuse))
solar_radius_degrees = 16 / 60
```

```
r = refraction(0.0, temperature_C=15.0,
pressure_mbar=1030.0)
```

Then, determine the time of the sunset

```
t, y = almanac.find_settings(location, sun, t0, t1,
horizon_degrees=-r + h.degrees - solar_radius_degrees)
h, m, s = t.utc.hour, t.utc.minute, t.utc.second
print(h,m,s)
[10.] [54.] [23.66874027]
```

Since the 'Isha (Isya') prayer occurs after sunset, it is necessary to perform a time-based loop starting from the moment of sunset and continuing until the solar altitude reaches 18 degrees of solar depression. The time loop operation is similar with the previous asar prayer time loop operation. First determine the solar altitude,

```
sun_astro = location.at(ts.utc(year, month, day, h,
m)).observe(sun)
sun_app = sun_astro.apparent()
sun_alt, sun_az, distance = sun_app.altaz()
print(sun_alt)
-00deg 58' 37.5"
```

To develop with hour time loop, with the rule of the loop the solar degree does not pass the altitude of – 17.

```
# Start with hour

test = 1
while True:

  # Calculate the Solar Altitude
  sun_astro = location.at(ts.utc(year, month, day,
  h+1, m, s)).observe(sun)
  sun_alt, _, _ = sun_astro.apparent().altaz() # Get
  the altitude of the sun
  isha_angle = 17
  elevation_correction = 0.0293 * math.sqrt(ele)
  isha_angle_actual = -isha_angle - elevation_correction

  if sun_alt.degrees >= 0:
    break
  if test > 24:
    break
```

```
    if sun_alt.degrees <= isha_angle_actual:
      break # Exit the loop if the solar altitude
      located below -18 degree
    h = h + 1
```

Then proceed with a minute-level time loop to more precisely determine the exact minute when the solar altitude is first reached.

```
  # Once the condition is met for hours, move to minutes
  h_isya = h - 1
  test = 1
  while True:
    # Calculate the Solar Altitude
    sun_astro = location.at(ts.utc(year, month, day,
    h_isya, m, s)).observe(sun)
    sun_alt, _, _ = sun_astro.apparent().altaz() # Get
    the altitude of the sun

    if sun_alt.degrees >= 0:
      break
    if test > 1440:
      break

    if sun_alt.degrees <= isha_angle_actual:
       break # Exit the loop if the solar altitude
       located below -18 degree
    m += 1

    # Increment time in minutes
  m_isya = m - 1
```

Proceed with a second-level time loop to more precisely determine the exact minute when the required Isya' depression degree is first reached.

```
  # Once the condition is met for minutes, move to
  seconds
  test = 1
  while True:
    # Calculate the Solar Altitude
    sun_astro = location.at(ts.utc(year, month, day,
    h_isya, m_isya, s)).observe(sun)
    sun_alt, _, _ = sun_astro.apparent().altaz() # Get
    the altitude of the sun
```

```
    if sun_alt.degrees >= 0:
      break
    if test > 86400:
      break

    if sun_alt.degrees <= isha_angle_actual:
       break # Exit the loop if the solar altitude
       located below -18 degree
    s += 1
    print(f'Sun Altitude {sun_alt.degrees}, {isha_angle_
    actual} at second {s}')

    # Increment time in seconds

s_isya = s
```

Then, we step back one second. The Isya' time, after the time loop, with timezone correction is

```
isya_time = (h_isya + (m_isya) / 60 + s_isya /
3600)+timezone
print(isya_time)
[19.05666667]
```

To correct this and ensure the resulting prayer time remains within a 24-hour format,

```
isya_time %= 24 # Ensure 24-hour clock format
print(isya_time)
[19.05666667]
```

To convert into time format,

```
isya_time = float(isya_time)
degrees = int(isya_time)
decimal_part = isya_time - degrees
minutes_total = decimal_part * 60
minutes = int(minutes_total)
seconds = round((minutes_total - minutes) * 60)

if sun_alt.degrees >= 0 or test >86400:
    isya = "Isya' Does Not Occur"
else:
    isya = f"Isya' Occurs at {degrees}° {minutes}'
    {seconds}""
```

```
print(isya)
Isya' Occurs at 19° 3' 25"
```

'Isya prayer time at Jakarta, Indonesia, where Latitude: 6.2088° South, Longitude: 106.8456° East, Time Zone: GMT+7, Elevation: 50 meters, Date: 21 July 2025, Assume 'Isya begins at a solar depression of −17 is at 19:3:24.

Exercise 2

Determine the 'Isya prayer time at Istanbul, Turkey, where Latitude: 41.0082° North, Longitude: 28.9784° East, Time Zone: GMT+3, Elevation: 40 meters, Date: 15 August 2025. Use a solar depression angle of −16° to mark the beginning of 'Isya prayer.

Exercise 3

Determine the 'Isya prayer time at Tokyo, Japan, with the following details, Latitude: 35.6762° North, Longitude: 139.6503° East, Elevation: 40 meters, Time Zone: GMT+9, Date: 16 June 2025. Assume 'Isya begins when the solar depression angle reaches 20° below the horizon.

SYURUK

Syuruk is the end of Subh prayer time. It is based on the timing of the sunrise. This originates from the prophetic saying;

> *...The beginning of the time for Fajr is when Fajr begins, and its end is when the sun rises.*

Sunrise from the Syuruk is the first visibility of the upper limb of the sun. This means that the upper limb of the sun must be fully visible from view. Similarly, the Maghrib refraction and elevation factor also need to be considered. To determine the time of the Syuruk, assuming the user is located at Sydney Tower Eye, with Latitude 33.8688° South, Longitude 151.2093° East, on 23 March 2025, where the time zone in effect is GMT + 11, with elevation of 350 meter. First, we need to import function that are needed for the calculation.

```
from skyfield.api import load
from skyfield.api import N, S, E, W, wgs84
from skyfield import almanac
import math
```

Put input into the variable

```
lat_location = -33.8688
long_location = 151.2093
timezone = 11
day = 23
month = 3
year = 2025
ele = 350
```

Load ephemeris.

```
ts = load.timescale()
eph = load('de440s.bsp')
planets = load('de440s.bsp')
earth = planets['earth']
sun = planets['sun']
```

Feed info of user location

```
location = earth + wgs84.latlon(lat_location, long_
location, elevation_m=ele)
```

Skyfield require range of calculate date, this is written as;

```
t0 = ts.utc(year, month, day-1)
t1 = ts.utc(year, month, day)
```

Since Syuruk occurs in the morning, we assume the calculation from the day before. Then, include the variable of refraction and location elevation

```
from skyfield.units import Angle
from numpy import arccos
from skyfield.earthlib import refraction

altitude_m = ele
earth_radius_m = 6378136.6
side_over_hypotenuse = earth_radius_m / (earth_
radius_m + altitude_m)
```

```
h = Angle(radians=-arccos(side_over_hypotenuse))
solar_radius_degrees = 16 / 60
r = refraction(0.0, temperature_C=15.0,
pressure_mbar=1030.0)
```

Then, determine the time of the sunrise

```
t, y = almanac.find_risings(location, sun, t0, t1,
horizon_degrees=-r + h.degrees - solar_radius_degrees)
h, m, s = t.utc.hour, t.utc.minute, t.utc.second
print(h,m,s)
[19.] [57.] [29.52933632]
```

The time of the sunset is 19:57:29, at UTC time. To convert into local time with GMT + 11.

```
syuruk_time = float(h + m / 60 + s / 3600 + timezone)
print(syuruk_time)
30.958202593423238
```

When calculating times near sunrise, there may be instances where the computed time exceeds 24 hours; the above example is the case. To correct this and ensure the resulting prayer time remains within a 24-hour format, the following code is used to normalize the time

```
syuruk_time %= 24 # Ensure 24-hour clock format
print(syuruk_time)
6.958202593423238
```

From this moment, we can convert into time format,

```
syuruk_time = float(syuruk_time)
degrees = int(syuruk_time)
decimal_part = syuruk_time - degrees
minutes_total = decimal_part * 60
minutes = int(minutes_total)
seconds = round((minutes_total - minutes) * 60)
sun_astro = location.at(ts.utc(year, month, day, h, m,
s)).observe(sun)
sun_alt, _, _ = sun_astro.apparent().altaz()
sun_alt, _, _ = sun_astro.apparent().altaz()
if sun_alt.degrees >= 0:
    syuruk = "Syuruk Does Not Occur"
else:
```

```
        syuruk = f"Syuruk Occurs at {degrees}° {minutes}'
        {seconds}"

    print(syuruk)
    Syuruk Occurs at 6° 57' 30
```

This means that syuruk at Sydney Tower Eye, with Latitude 33.8688° South, Longitude 151.2093° East, on 23 March 2025, where the timezone in effect is GMT + 11, with elevation of 350 meters occurs at 06:57:30.

Exercise 1

Determine the Syuruk (sunrise end) time on 15 July 2025 for a user located in central Tokyo, Japan, with the following location details:

- Latitude: 35.6762° North
- Longitude: 139.6503° East
- Elevation: 40 meters
- Time Zone: GMT +9

Exercise 2

Calculate the Syuruk time for Cape Town, South Africa, on 10 October 2025. Use the following coordinates:

- Latitude: 33.9249° South
- Longitude: 18.4241° East
- Elevation: 15 meters
- Time Zone: GMT +2

Exercise 3

On 1 January 2025, determine the Syuruk prayer time for a user located in Rio de Janeiro, Brazil:

- Latitude: 22.9068° South
- Longitude: 43.1729° West
- Elevation: 5 meters
- Time Zone: GMT –3

Exercise 4

Compute the Syuruk time for Berlin, Germany, on 28 May 2025 using the following details:

- Latitude: 52.5200° North
- Longitude: 13.4050° East
- Elevation: 34 meters
- Time Zone: GMT +2 (Daylight Saving Time in effect)

SUBH

The beginning of Subh' prayer time is based on the position of the sky condition before sunrise. This is based on the prophetic saying.

> *...Then he prayed Fajr when Fajr (dawn) began...*
> *...then he prayed Subh when the land glowed*

The appearance of twilight from the sky is caused by the sun's continued ascent below the horizon before sunrise. At a certain angle, known as the solar depression angle, the sun's rays begin to be refracted by the atmosphere in a way that illuminates the night sky. When this critical angle is reached, the first dim light appears horizontally in the sky, marking the beginning of the Subh prayer time. However, Islamic scholars differ in their opinions regarding the exact degree of solar depression that signifies the start of Subh. While many adopt angles between 15° and 20° below the horizon, there is no unanimous agreement, and various regions apply different standards based on jurisprudential reasoning and observational studies. To determine the Subh prayer time in Bangkok, where Latitude 13.7563° North, Longitude 100.5018° East and Time Zone GMT+7, with elevation of 150 meter on 16 June 2025 for 18 degrees of solar depression during Subh, we need to import function that are needed for the calculation.

```
from skyfield.api import load
from skyfield.api import N, S, E, W, wgs84
from skyfield import almanac
import math
```

Put input into the variable

```
lat_location = 13.7563
long_location = 100.5018
timezone = 7
day = 16
month = 6
year = 2025
ele = 150
```

Load ephemeris.

```
ts = load.timescale()
eph = load('de440s.bsp')
planets = load('de440s.bsp')
earth = planets['earth']
sun = planets['sun']
```

Feed info of user location

```
location = earth + wgs84.latlon(lat_location, long_
location, elevation_m=ele)
```

Skyfield require range of calculate date, this is written as;

```
t0 = ts.utc(year, month, day-1)
t1 = ts.utc(year, month, day)
```

Then, include the variable of refraction and location elevation

```
from skyfield.units import Angle
from numpy import arccos
from skyfield.earthlib import refraction

altitude_m = ele
earth_radius_m = 6378136.6
side_over_hypotenuse = earth_radius_m / (earth_
radius_m + altitude_m)
h = Angle(radians=-arccos(side_over_hypotenuse))
solar_radius_degrees = 16 / 60
r = refraction(0.0, temperature_C=15.0,
pressure_mbar=1030.0)
```

Then, determine the time of the sunrise

```
t, y = almanac.find_risings(location, sun, t0, t1,
horizon_degrees=-r + h.degrees - solar_radius_degrees)
h, m, s = t.utc.hour, t.utc.minute, t.utc.second
print(h,m,s)
[22.] [48.] [48.62638382]
```

Since the Subh prayer occurs before sunrise, it is necessary to perform a time-based loop starting from the moment of sunrise and backtrack until the solar altitude reaches 18 degrees of solar depression. The time loop operation is similar with the previous Asar prayer time loop operation. First determine the solar altitude,

```
sun_astro = location.at(ts.utc(year, month, day, h,
m)).observe(sun)
sun_app = sun_astro.apparent()
sun_alt, sun_az, distance = sun_app.altaz()
print(sun_alt)
-01deg 27' 11.4"
```

To develop with hour time loop, with the rule of the loop the solar degree does not pass the altitude of – 18.

```
# Start with hour
test = 1
while True:

  # Calculate the Solar Altitude
  sun_astro = location.at(ts.utc(year, month, day, h,
  m, s)).observe(sun)
  sun_alt, _, _ = sun_astro.apparent().altaz() # Get
  the altitude of the sun
  subh_angle = 18
  elevation_correction = 0.0293 * math.sqrt(ele)
  subh_angle_actual = -subh_angle
  - elevation_correction

  if sun_alt.degrees >= 0:
    break
  if test > 24:
    break
```

```
    if sun_alt.degrees <= subh_angle_actual:
       break # Exit the loop if the solar altitude
       located below -18 degree
    h -= 1
    print(f'Sun Altitude {sun_alt.degrees}, at hour
    {h}')

Sun Altitude [-1.27408764], at hour [21.]
Sun Altitude [-14.3007989], at hour [20.]
```

In the above code, if sun_alt.degrees <= subh_angle_actual is used to determine whether the solar altitude has reach below 18 degree of depression. From the output, the sun altitude reach angle below −18 degrees at hour 20. Then proceed with a minute-level time loop to more precisely determine the exact minute when the solar altitude is first reached.

```
# Once the condition is met for hours, move to minutes
h_subh = h + 1
test = 1
while True:
  # Calculate the Solar Altitude
  sun_astro = location.at(ts.utc(year, month, day,
  h_subh, m, s)).observe(sun)
  sun_alt, _, _ = sun_astro.apparent().altaz() # Get
  the altitude of the sun

  if sun_alt.degrees >= 0:
    break
  if test > 1440:
    break

  if sun_alt.degrees <= subh_angle_actual:
        break # Exit the loop if the solar altitude
        located below
  m -= 1
  print(f'Sun Altitude {sun_alt.degrees}, at minute
  {m}')

  # Increment time in minutes
m_subh = m + 1

Sun Altitude [-14.3007989], at minute [47.]
Sun Altitude [-14.51317791], at minute [46.]
Sun Altitude [-14.72537202], at minute [45.]
Sun Altitude [-14.93737954], at minute [44.]
```

```
Sun Altitude [-15.14919881], at minute [43.]
Sun Altitude [-15.36082811], at minute [42.]
Sun Altitude [-15.57226573], at minute [41.]
Sun Altitude [-15.78350995], at minute [40.]
Sun Altitude [-15.994559], at minute [39.]
Sun Altitude [-16.20541111], at minute [38.]
Sun Altitude [-16.41606451], at minute [37.]
Sun Altitude [-16.62651737], at minute [36.]
Sun Altitude [-16.83676789], at minute [35.]
Sun Altitude [-17.04681421], at minute [34.]
Sun Altitude [-17.25665447], at minute [33.]
Sun Altitude [-17.4662868], at minute [32.]
Sun Altitude [-17.67570929], at minute [31.]
Sun Altitude [-17.88492002], at minute [30.]
Sun Altitude [-18.09391709], at minute [29.]
Sun Altitude [-18.30269847], at minute [28.]
```

From the printed output above, we can see that the sun altitude reaches below the required Subh depression degree at minute 28. Therefore, we step back one minute and proceed with a second-level time loop to more precisely determine the exact minute when the required Subh depression degree is first reached.

```
# Once the condition is met for minutes, move to
seconds
test = 1
while True:
  # Calculate the Solar Altitude
  sun_astro = location.at(ts.utc(year, month, day,
  h_subh, m_subh, s)).observe(sun)
  sun_alt, _, _ = sun_astro.apparent().altaz() # Get
  the altitude of the sun
  if sun_alt.degrees >= 0:
    break
  if test > 86400:
    break

  if sun_alt.degrees <= subh_angle_actual:
          break # Exit the loop if the solar altitude
          located below
  s -= 1
  print(f'Sun Altitude {sun_alt.degrees}, at second
  {s}')

  # Increment time in seconds

s_subh = s +1
```

```
Sun Altitude [-18.30269847], at second [47.62638382]
Sun Altitude [-18.30617632], at second [46.62638382]
Sun Altitude [-18.30965412], at second [45.62638382]
Sun Altitude [-18.31313185], at second [44.62638382]
Sun Altitude [-18.31660952], at second [43.62638382]
Sun Altitude [-18.32008713], at second [42.62638382]
Sun Altitude [-18.32356467], at second [41.62638382]
Sun Altitude [-18.32704216], at second [40.62638382]
Sun Altitude [-18.33051959], at second [39.62638382]
Sun Altitude [-18.33399696], at second [38.62638382]
Sun Altitude [-18.33747427], at second [37.62638382]
Sun Altitude [-18.34095151], at second [36.62638382]
Sun Altitude [-18.3444287], at second [35.62638382]
Sun Altitude [-18.34790582], at second [34.62638382]
Sun Altitude [-18.35138289], at second [33.62638382]
Sun Altitude [-18.35485989], at second [32.62638382]
Sun Altitude [-18.35833683], at second [31.62638382]
```

From the printed output above, we can see that the solar altitude exceeds the required Isy' Solar Depression at second 31. Therefore, we step back one second. The Isya' time, after the time loop, with timezone correction is

```
subh_time = (h_subh + (m_subh) / 60 + s_subh /
3600)+timezone
print(subh_time)
[28.4925]
```

To correct this and ensure the resulting prayer time remains within a 24-hour format,

```
subh_time %= 24 # Ensure 24-hour clock format
print(subh_time)
[4.4925]
```

To convert into time format,

```
subh_time = float(subh_time)
degrees = int(subh_time)
decimal_part = subh_time - degrees
minutes_total = decimal_part * 60
minutes = int(minutes_total)
seconds = round((minutes_total - minutes) * 60)

if sun_alt.degrees >= 0 or test >86400:
    subh = "Subuh Does Not Occur"
else:
```

```
    subh = f"Subuh Occurs at {degrees}° {minutes}'
    {seconds}""

print(subh)
Subuh Occurs at 4° 29' 33"
```

Subh prayer time at Bangkok where Latitude 13.7563° North, Longitude 100.5018° East and Time Zone GMT+7, with elevation of 150 meter on 16 June 2025 for 18 degree of solar depression is at 04:31:16.

Exercise 1

Determine the Subh prayer time at Jakarta, Indonesia, where Latitude: 6.2088° South, Longitude: 106.8456° East, Time Zone: GMT+7, Elevation: 50 meters, Date: 21 July 2025, Assuming Subh begins at a solar depression of –17°. First, we need to import function that are needed for the calculation.

```
from skyfield.api import load
from skyfield.api import N, S, E, W, wgs84
from skyfield import almanac
import math
```

Put input into the variable

```
lat_location = -6.2088
long_location = 106.8456
timezone = 7
day = 21
month = 7
year = 2025
ele = 50
```

Load ephemeris.

```
ts = load.timescale()
eph = load('de440s.bsp')
planets = load('de440s.bsp')
earth = planets['earth']
sun = planets['sun']
```

Feed info of user location

```
location = earth + wgs84.latlon(lat_location, long_
location, elevation_m=ele)
```

Skyfield require range of calculate date, this is written as;

```
t0 = ts.utc(year, month, day-1)
t1 = ts.utc(year, month, day)
```

Then, include the variable of refraction and location elevation

```
from skyfield.units import Angle
from numpy import arccos
from skyfield.earthlib import refraction

altitude_m = ele
earth_radius_m = 6378136.6
side_over_hypotenuse = earth_radius_m / (earth_
radius_m + altitude_m)
h = Angle(radians=-arccos(side_over_hypotenuse))
solar_radius_degrees = 16 / 60
r = refraction(0.0, temperature_C=15.0,
pressure_mbar=1030.0)
```

Then, determine the time of the sunrise

```
t, y = almanac.find_risings(location, sun, t0, t1,
horizon_degrees=-r + h.degrees - solar_radius_degrees)
h, m, s = t.utc.hour, t.utc.minute, t.utc.second
```

Since the Subh prayer occurs before sunrise, it is necessary to perform a time-based loop starting from the moment of sunrise and backtrack until the solar altitude reaches 17 degrees of solar depression. The time loop operation is similar with the previous Asar prayer time loop operation. First determine the solar altitude,

```
sun_astro = location.at(ts.utc(year, month, day, h,
m)).observe(sun)
sun_app = sun_astro.apparent()
sun_alt, sun_az, distance = sun_app.altaz()
```

To develop with hour time loop, with the rule of the loop the solar degree does not pass the altitude of – 17.

```
# Start with hour

test = 1
while True:

  # Calculate the Solar Altitude
```

```
    sun_astro = location.at(ts.utc(year, month, day, h,
    m, s)).observe(sun)
    sun_alt, _, _ = sun_astro.apparent().altaz() # Get
    the altitude of the sun
    subh_angle = 17
    elevation_correction = 0.0293 * math.sqrt(ele)
    subh_angle_actual = -subh_angle
    - elevation_correction
    if sun_alt.degrees >= 0:
      break
    if test > 24:
      break

    if sun_alt.degrees <= subh_angle_actual:
       break # Exit the loop if the solar altitude
       located below -18 degree
    h -= 1
```

In the above code, if sun_alt.degrees <= subh_angle_actual:is used to determine whether the solar altitude has reached below, then proceed with a minute-level time loop to more precisely determine the exact minute when the solar altitude is first reached.

```
  # Once the condition is met for hours, move to minutes
  h_subh = h + 1
  test =1
  while True:
    # Calculate the Solar Altitude
    sun_astro = location.at(ts.utc(year, month, day,
    h_subh, m, s)).observe(sun)
    sun_alt, _, _ = sun_astro.apparent().altaz() # Get
    the altitude of the sun
    if sun_alt.degrees >= 0:
      break
    if test > 1440:
      break

    if sun_alt.degrees <= subh_angle_actual:

       break # Exit the loop if the solar altitude
       located below -18 degree
    m -= 1

    # Increment time in minutes
  m_subh = m + 1
```

Then proceed with a second-level time loop to more precisely determine the exact minute when the required Subh depression degree is first reached.

```
# Once the condition is met for minutes, move to
seconds
test = 1
while True:
  # Calculate the Solar Altitude
  sun_astro = location.at(ts.utc(year, month, day,
  h_subh, m_subh, s)).observe(sun)
  sun_alt, _, _ = sun_astro.apparent().altaz() # Get
  the altitude of the sun
  if sun_alt.degrees >= 0:
    break
  if test > 86400:
    break

  if sun_alt.degrees <= subh_angle_actual:
     break # Exit the loop if the solar altitude
     located below -18 degree
  s -= 1

  # Increment time in seconds

  s_subh = s +1
```

The Isya' time, after the time loop, with timezone correction is

```
subh_time = (h_subh + (m_subh) / 60 + s_subh /
3600)+timezone
print(subh_time)
[28.91295064]
```

To correct this and ensure the resulting prayer time remains within a 24-hour format,

```
subh_time %= 24 # Ensure 24-hour clock format
print(subh_time)
4.91295064
```

To convert into time format,

```
subh_time = float(subh_time)
degrees = int(subh_time)
```

```
decimal_part = subh_time - degrees
minutes_total = decimal_part * 60
minutes = int(minutes_total)
seconds = round((minutes_total - minutes) * 60)

if sun_alt.degrees >= 0 or test >86400:
    subh = "Subuh Does Not Occur"
else:
    subh = f"Subuh Occurs at {degrees}° {minutes}'
    {seconds}""

print(subh)
Subuh Occurs at 4° 54' 47"
```

Subh prayer time at Jakarta, Indonesia, where Latitude: 6.2088° South, Longitude: 106.8456° East, Time Zone: GMT+7, Elevation: 50 meters, Date: 21 July 2025, Assuming Subh begins at a solar depression of –17° is at 04:54:47.

Exercise 2

Determine the Subh (Fajr) prayer time at Khartoum, Sudan on 21 July 2025, assuming Subh begins when the sun reaches –15° below the horizon. Location details:

- Latitude: 15.5007° North
- Longitude: 32.5599° East
- Elevation: 380 meters
- Time Zone: GMT+2

Exercise 3

Calculate the Subh prayer time in Manila, Philippines for 21 July 2025, using a –19° solar depression angle. Location details:

- Latitude: 14.5995° North
- Longitude: 120.9842° East
- Elevation: 16 meters
- Time Zone: GMT+8

Exercise 4

Compute the Subh prayer time at San Salvador, El Salvador on 21 July 2025, based on a –20° solar depression. Location details:

- Latitude: 13.6929° North
- Longitude: –89.2182° West
- Elevation: 658 meters
- Time Zone: GMT–6

8 Moonsighting Observation Data Computation

The determination of new Hijri months is based on the position of the moon during 29th day of a Hijri month (Mustapha et al., 2024). Either the new Hijri month is determined using moonsighting, astronomical calculations, or lunar crescent visibility criteria; knowing the geometric position of the moon is vital. In the past, the endeavor to calculate the position of the moon during 29th day of Hijri month is an arduous task, and only selected person can carry out the task. Nowadays, due to the invention of Python programming language and extensive libraries, the task of calculating the position of the moon can be conducted with a few lines of code. In this case, the task of calculating the position of the moon can be calculated using Skyfield.

Let us say, user observation site is Kigali, Rwanda, with Latitude 1.9577° South, Longitude: 30.1127° East, Elevation: 1,567 meters and Time Zone: GMT+2. Determine the geometrical position of the moon during for the Hijri month observation on 27 May 2025. First, install skyfield, and other related libraries.

```
pip install skyfield numpy tabulate
```

Import-related functions

```
from skyfield import almanac
from skyfield.api import Topos,load
from skyfield import api
import numpy as np
from skyfield.api import N, S, E, W, load, wgs84
from skyfield.api import Topos, load,
Angle,GREGORIAN_START
import math
```

DOI: 10.1201/9781003649120-8

```
from scipy.ndimage import rotate
import calendar
from tabulate import tabulate
from matplotlib.patches import Arc
import matplotlib.pyplot as plt
from matplotlib.colors import LinearSegmentedColormap
import matplotlib.image as mpimg
from matplotlib.offsetbox import OffsetImage,
AnnotationBbox
```

Load ephemeris and function

```
planets = load('de440s.bsp')
earth = planets['earth']
sun = planets['sun']
moon = planets['moon']
h_maghrib = 0
m_maghrib = 0
ts = load.timescale()
eph = api.load('de440s.bsp')
```

The moon is loaded since we calculate the position of the moon. h_maghrib = 0, m_maghrib = 0

is declared to mitigate some issue later. Then, input the related variables.

```
lat_location = -1.9577
long_location = 30.1127
timezone = 8
year = 2025
month = 3
day = 30
ele = 100
```

Input the location info and refraction and elevation parameter.

```
location = earth + wgs84.latlon(lat_location, long_
location, elevation_m=0)
from skyfield.units import Angle
from numpy import arccos
from skyfield.earthlib import refraction
altitude_m = ele
earth_radius_m = 6378136.6
side_over_hypotenuse = earth_radius_m / (earth_
radius_m + altitude_m)
h = Angle(radians=-arccos(side_over_hypotenuse))
```

```
solar_radius_degrees = 16 / 60
r = refraction(0.0, temperature_C=15.0,
pressure_mbar=1030.0)
```

Determine the time of sunset.

```
t, y = almanac.find_settings(location, sun, t0, t1,
horizon_degrees=-r + h.degrees - solar_radius_degrees)
h_sunset, m_sunset, s_sunset = t.utc.hour, t.utc.minu
te, t.utc.second
```

In moonsighting calculations, the sunset time is stored using the variables h_sunset, m_sunset, s_sunset. This distinction is made to clearly differentiate it from the moonset time, which will be calculated separately. Then convert sunset time in time format

```
sunset_time = float(h_sunset + m_sunset / 60 + s_
sunset / 3600 + timezone)
sunset_time %= 24 # Ensure 24-hour clock format
sunset_time = float(sunset_time)
degrees = int(sunset_time)
decimal_part = sunset_time - degrees
minutes_total = decimal_part * 60
minutes = int(minutes_total)
seconds = round((minutes_total - minutes) * 60)
sunset = f"{degrees}° {minutes}' {seconds}""
print(sunset)

18° 7' 58"
```

We can combine the result into a variable using an f-string, as shown above. In this example, the result is stored in the sunset variable and printed in degree-minute-second (DMS) format. This corresponds to 18:07:58, which indicates the time of sunset. Once the sunset time is determined, we can proceed to calculate the moonset time, which will follow a similar method.

```
t, y = almanac.find_settings(location, moon, t0, t1,
horizon_degrees=-r + h.degrees - solar_radius_degrees)
h_moonset, m_moonset, s_moonset = t.utc.hour, t.utc
.minute, t.utc.second
```

Note the changes from sun to moon in comparison to the previous code. This is because the previous code is to calculate the sun position, while this code is used to calculate the moon position. Then convert moonset time in time format

```
moonset_time = float(h_moonset + m_moonset / 60 +
s_moonset / 3600 + timezone)
moonset_time %= 24 # Ensure 24-hour clock format
moonset_time = float(moonset_time)
degrees = int(moonset_time)
decimal_part = moonset_time - degrees
minutes_total = decimal_part * 60
minutes = int(minutes_total)
seconds = round((minutes_total - minutes) * 60)
moonset = f"{degrees}° {minutes}′ {seconds}″"
print(moonset)
19° 4′ 59″
```

Therefore, the moonset occurs at 19:04:59. Then the value of lag time, the difference of time between moonset and sunset, can be extracted.

```
lag_time = abs(moonset_time - sunset_time)
print(lag_time)
0.9504272807035683
```

To calculate lag time, subtract **moonset_time** and **sunset_time**. Then make sure the result is always positive; use the function abs() inside the subtraction operation. The result is in decimal format; to convert into time format,

```
degrees = int(lag_time)
decimal_part = lag_time - degrees
minutes_total = decimal_part * 60
minutes = int(minutes_total)
seconds = round((minutes_total - minutes) * 60)
lagtime= f"{degrees}° {minutes}′ {seconds}″"
print(lagtime)
0° 57′ 2″
```

The lag time between moonset and sunset is 57 minute and 2 seconds. Next is to determine the geometrical position of the moon. The geometrical position of the moon is calculated using a topocentric reference. This means that the position of the moon is calculated in reflect to the user position on the surface of the Earth. The calculation is conducted based on the position of the moon during sunset. To perform the calculation, first we calculate the altitude of the moon during sunset.

```
moon_astro = location.at(ts.utc(year, month, day,
h_sunset, m_sunset)).observe(moon)
```

```
moon_app = moon_astro.apparent()
moon_alt, moon_az, distance = moon_app.altaz()
print(moon_alt)
12deg 27′ 16.3″
```

The code structure used to calculate moonset is like the code for calculating prayer times, such as sunset. The only difference lies in the celestial object being observed, the sun for prayer times, and the moon for moonsighting. In this case, the same variables as h_sunset and m_sunset can be reused for determining the moon's position during sunset. The moon altitude during sunset is 12 deg 27' 16.3". This format is readable for humans but not suitable for numerical calculations. To perform further computations, like determining moon visibility, we need the altitude in decimal degrees. To do this, simply access the .degrees attribute of the moon_alt variable:

```
print(moon_alt.degrees)
[12.4545324]
```

The next important parameter for new moon sighting is arc of vision. The arc of vision is the altitude difference between the sun's altitude and moon's altitude. First, determine the sun's altitude during sunset.

```
sun_astro = location.at(ts.utc(year, month, day,
h_sunset, m_sunset)).observe(moon)
sun_app = sun_astro.apparent()
sun_alt, sun_az, distance = sun_app.altaz()
print(sun_alt.degrees)
[-0.92325118]
```

Next, the difference between sun-moon altitude or Arc of Vision,

```
arc_of_vision = abs(moon_alt.degrees - sun_alt.
degrees)
print(arc_of_vision)
[13.37778358]
```

The arc of vision is 13.3778358. Next is the arc of light. The arc of light is the angle of separation between sun and moon, or the elongation angle. The arc of light can be determined with code,

```
arc_of_light = sun_app.separation_from(moon_app)
print(arc_of_light.degrees)
[16.57460685]
```

The arc of light is 13.3778358. Next is the difference in azimuth. The difference in azimuth is the azimuthal difference between sun azimuth and moon azimuth. The code to determine the difference in azimuth is as follows

```
difference_azimuth =
abs(moon_az.degrees-sun_az.degrees)
print(difference_azimuth)
[9.85833904]
```

The difference in azimuth between the sun and the moon is 9.85833904°. The final parameter to consider is the moon age, which refers to the elapsed time between the moment of conjunction (new moon) and the sunset at the observation location. This age is a crucial factor in evaluating the likelihood of crescent visibility. To determine moon age, first determine the Julian date of the sunset time,

```
t, y = almanac.find_settings(location, moon, t0, t1,
horizon_degrees=-r + h.degrees - solar_radius_degrees)
jd_sunset = t.tt

jd_sunset= jd_sunset[0]
print(jd_sunset)
2460760.037716627
```

The code is similar with find sunset code, however the output is changed to t.tt, as it is in Julian date format. Next, determine the time of moon conjunction using almanac function,

```
t0 = ts.utc((year), (month), (day-5))
t1 = ts.utc((year), (month), (day+5))
f = almanac.oppositions_conjunctions(eph, eph['Moon'])
t, y = almanac.find_discrete(t0, t1, f)

for ti, yi in zip(t, y):
   if yi == 1:
       jd_moon_conjuction = format(ti.tt)
   else:
       None
jd_moon_conjuction= float(jd_moon_conjuction)
```

The time of moon conjunction is computed using almanac.oppositions_conjunctions function. This function can determine the timing of conjunction or opposition, with yi == 1 for conjunction, and yi == 0 for moon opposition or full moon. –5 and +5 is performed at t0 and t1 since the conjunction usually takes a 0 to a few days before the moon observation. Then, the moon age is computed as the Julian date difference between sunset and moon conjunction.

```
moonage = (jd_sunset -jd_moon_conjuction)*24
print(moonage)
```

There! We got the geometric position of the moon during observation.

Exercise 1

An observer in Istanbul, Turkey, located at 41.0082° North latitude and 28.9784° East longitude, at an elevation of 39 meters and under the GMT+3 time zone, intends to sight the new moon that would indicate the beginning of the month of Shawal. The observation is to be made on 29 March 2025, which corresponds to the 29th day of Ramadan 1446H. Determine the geometrical position of the moon during this date and time.

Exercise 2

In Cape Town, South Africa, an observer situated at 33.9249° South latitude and 18.4241° East longitude, at an elevation of 15 meters and within the GMT+2 time zone, will attempt to observe the crescent moon on 6 June 2025. This date aligns with the 29th day of Zulkaedah 1446H and is critical for determining the beginning of the month of Zulhijjah. Calculate the geometrical position of the moon at this location and date.

Exercise 3

Assume a user in Jakarta, Indonesia, with coordinates of 6.2088° South latitude and 106.8456° East longitude, observes the moon from an elevation of 50 meters above sea level. The local time zone is GMT+7. The observation is planned for 29 March 2025, corresponding to 29 Ramadhan 1446H. The task is to determine the geometrical position of the moon for Shawwal crescent sighting on that evening.

Exercise 4

In Los Angeles, USA, an observer located at 34.0522° North latitude and 118.2437° West longitude, at an elevation of 71 meters and within the GMT–7 time zone, intends to observe the moon on 6 June 2025. This corresponds to the 29th day of Zulkaedah 1446H. Determine the geometrical position of the moon during sunset at this location to evaluate the possibility of crescent sighting for Zulhijjah.

9 Qiblah Compass Visualization

The determination of Rashdul Kiblat is universally suitable for all days in a year. There are times when the Rashdul Kiblat is not applicable for a particular day or time. This is due to the solar position. Therefore, another alternative is to determine the Qibla is based on the degree of solar azimuth. So, to create the visualization of the Qibla compass, it must be based on the solar azimuth degree. The solar azimuth degree can be calculated using Skyfield, while the Qibla direction can be calculated using the previous formula; therefore, the angle between the solar azimuth degree and Qibla direction can be determined using the subtraction between the solar azimuth degree and Qibla direction degree. This angle of subtraction can be visualized.

Visualization of the angle between the sun azimuth and the direction of the Qibla can be determined using matplotlib function of polar plot. Matplotlib is a visualization function that is embedded with Python. Matplotlib can be used to create bar graphs, line graph, and many more graphs. Matplotlib can visualize the angle of difference between solar azimuth and Qibla direction using polar plot. Polar plot is plot of magnitude $|G(j\omega)H(j\omega)|$ versus phase angle $\angle(G(j\omega)H(j\omega))$ in polar coordinates, and the value of frequency i.e. ω is varied from 0 to ∞. In polar plot magnitude of transfer function is plotted to distance from origin and phase angle is plotted from positive real x-axis. Polar plot is used in Nyquist plot to determine the stability of closed loop control system from its open loop frequency response. The polar plot has the angle scale from 0 degree to 360 degree. This would be perfect for the determination of Qibla direction from the position of the solar azimuth.

DOI: 10.1201/9781003649120-9

QIBLA DIRECTION VISUALIZATION ON POLAR PLOT

The instruction to use Polar Plots for Qibla direction visualization in Python is

```
fig, ax = plt.subplots(subplot_kw={'projection':
'polar'}, figsize=(8, 8))
where
fig: The figure object, which is the overall container
for the plot.
ax: The axes object, which represents the polar
subplot.
subplot_kw={'projection': 'polar'}: Configures the
axes to have a polar projection. The output for the
code
import matplotlib.pyplot as plt
# Create the polar plot
fig, ax = plt.subplots(subplot_kw={'projection':
'polar'}, figsize=(8, 8))
```

This code will output as shown in Figure 9.1.

Notice that the direction of 0 degree start from the right side of the compass. To rearrange the polar plot, with 0 degree located at the forward/upside of the compass, the code is

```
import matplotlib.pyplot as plt
# Create the polar plot
fig, ax = plt.subplots(subplot_kw={'projection':
'polar'}, figsize=(8, 8))
ax.set_theta_direction(-1) # Set clockwise direction
ax.set_theta_zero_location('N') # Set 0° to the top
(North)
```

Where ax.set_theha_zero_location ('N') will set the top as the starting location for the compass, which is the **North**. While ax.set_theta_direction(−1), will set the direction of the compass to be in clockwise direction, following the actual compass direction.

The direction of the Qibla follows the previous Qibla equation; let's say the direction of the Qibla is 291. In the code, this can be expressed as

```
qibla_direction_deg = 291
```

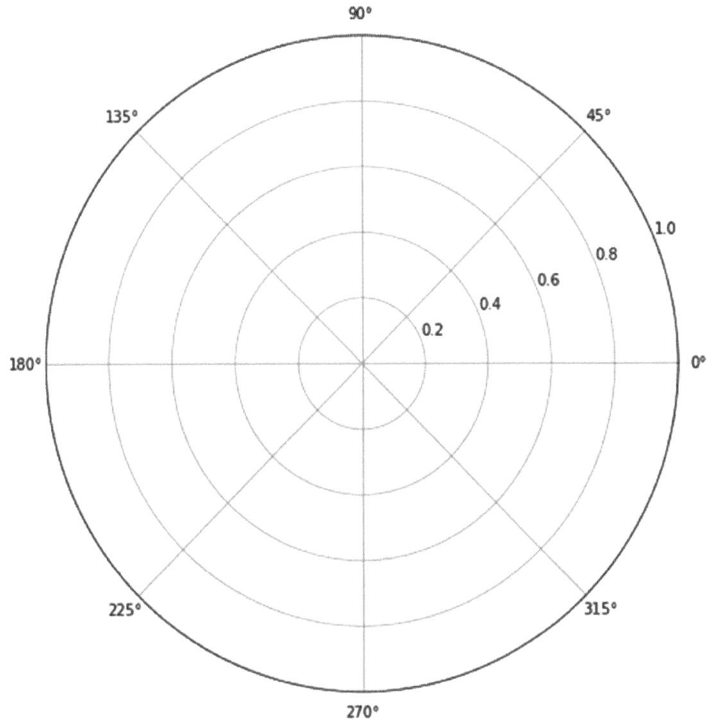

FIGURE 9.1 Generic polar plot.

The direction of the Qibla in the polar plot is read in radians form; therefore, to convert into radians form, the code is expressed as

```
qibla_direction_rad = np.deg2rad(qibla_direction_deg)
```

this will convert the angle degree to radians form. After that, to visualize the Qibla direction in the polar plot, the code is as follows

```
ax.plot([0, qibla_direction_rad], [0, 1], label='Qibla
Direction', color='blue', linewidth=2)
```

where,

```
ax.plot:
```

This is a method to plot data on the specified axes (ax), in this case, a polar axis created earlier.

`Arguments: [0, qibla_direction_rad] and [0, 1]:`

- [0, qibla_direction_rad]: This is the radial angle data in radians. The line starts at 0 radians (the center) and extends to qibla_direction_rad, which represents the direction of the Qibla in polar coordinates.
- [0, 1]: This is the radius data. The line starts at a radius of 0 (the center) and extends outward to a radius of 1.

`label='Qibla Direction':`

- This provides a label for the line, which can be displayed in a legend if added to the plot.

`color='blue':`

- Sets the color of the line to blue.

`linewidth=2:`

- Specifies the thickness of the line as 2 units.

The output for this code is shown in Figure 9.2.

Exercise 1: Visualize a Qibla Direction Using Location Latitude of 39.12 North, and Longitude of 80.11 East

First, determine the Qibla direction. To do this, first we insert the variable required for the calculation.

```
φ_Location = 39.12
λ_Location = 80.11
φ_Kaabah = 21.4225
λ_Kaabah = 39.8262
Difference_Longitude = abs(λ_Location-λ_Kaabah)
```

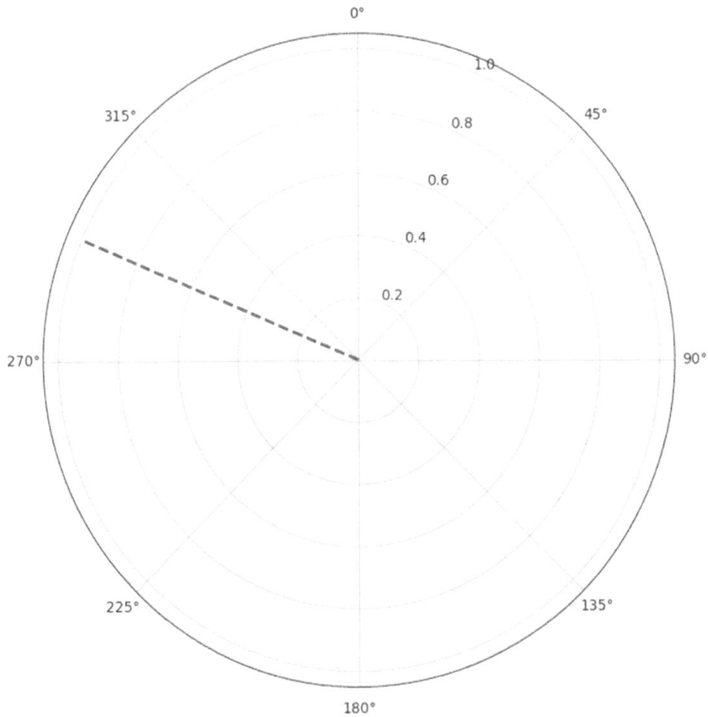

FIGURE 9.2 Polar plot with Qibla location.

Then, we performed the calculation.

```
#Calculation of Qibla Direction
import math

A = math.sin(math.radians(abs(Difference_Longitude)))
B = math.cos(math.radians(φ_Location))*math.tan(math.r
    adians(φ_Kaabah))
C = math.sin(math.radians(φ_Location)) * math.cos(m
    ath.radians(Difference_Longitude))
D = A/(B-C)
θ = math.degrees(math.atan(D))

#Determine the Azimuth of the Qibla
if Difference_Longitude > 180:
   delta_λ = 360 - Difference_Longitude
```

```
else:
    delta_λ = Difference_Longitude

if θ > 0:
    if λ_Location > λ_Kaabah:
        quadrant = "UB" # Utara Barat
    elif λ_Location <= λ_Kaabah:
        quadrant = "UT" # Utara Timur
    elif λ_Location < 0:
        if c >= 180:
          quadrant = "UB"
        else:
          quadrant = "UT"
elif θ < 0:
    if λ_Location > λ_Kaabah:
        quadrant = "SB" # Selatan Barat
    elif λ_Location <= λ_Kaabah:
        quadrant = "ST" # Selatan Timur
    elif λ_Location < 0:
        if c >= 180:
          quadrant = "SB"
        else:
          quadrant = "ST"

if quadrant == "UB":
    azimuth_kiblat = 360 - θ
elif quadrant == "SB":
    azimuth_kiblat = 180 - θ
elif quadrant == "UT":
    azimuth_kiblat = θ
elif quadrant == "ST":
    azimuth_kiblat = 180 + θ
# To Convert in Degree Form

degrees = int(azimuth_kiblat)
decimal_part = azimuth_kiblat - degrees
minutes_total = decimal_part * 60
minutes = int(minutes_total)
seconds = round((minutes_total - minutes) * 60)
print(f'The azimuth of the Qibla for Location with
coordinate {φ_Location} Latitude, {λ_Location}
Longitude, is {degrees}° {minutes}' {seconds}"')
```

```
The azimuth of the Qibla for Location with coordinate
39.12 Latitude, 80.11 Longitude, is 254° 41′ 48″
```

The Qibla direction for the given location is 254° 41′ 48″. Next, is to create the polar plot. First, create an empty polar plot.

```
import matplotlib.pyplot as plt
import numpy as np
fig, ax = plt.subplots(subplot_kw={'projection':
'polar'}, figsize=(8, 8))
```

The above code will result in an empty polar plot. And then, to align the top of the polar plot as North, or 0 degree.

```
import matplotlib.pyplot as plt
import numpy as np

fig, ax = plt.subplots(subplot_kw={'projection':
'polar'}, figsize=(8, 8))
ax.set_theta_direction(-1) # Set clockwise rotation
ax.set_theta_zero_location('N') # Set 0° (North) at
the top
```

Ok now, the top of the polar plot is 0 degree, similar with the alignment that we usually found on magnetic compass. Then is to label the Qibla direction based on the Qibla direction calculated before.

```
import matplotlib.pyplot as plt
import numpy as np

fig, ax = plt.subplots(subplot_kw={'projection':
'polar'}, figsize=(8, 8))
ax.set_theta_direction(-1) # Set clockwise rotation
ax.set_theta_zero_location('N') # Set 0° (North) at
the top

# Convert Qibla direction to radians for plotting
qibla_direction_rad = np.deg2rad(azimuth_kiblat)
ax.plot([0, qibla_direction_rad], [0, 1], label=
f'Qibla: {azimuth_kiblat:.2f}°', color='blue',
linewidth=2)
```

This will label the Qibla direction (Figure 9.3).

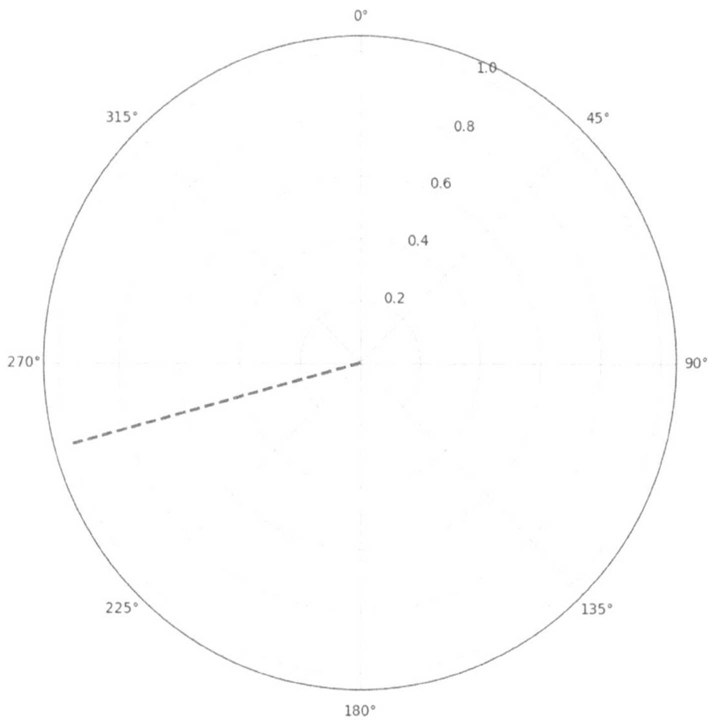

FIGURE 9.3 Polar plot with Qibla direction.

SUN AZIMUTH VISUALIZATION ON POLAR PLOT

After the Qibla direction can be computed and visualized in the polar plot, now it is for the solar azimuth computation. The solar azimuth computation can be performed using Skyfield. To calculate the sun azimuth at any given time and location, is as follows:

```
from skyfield.api import load
from skyfield.api import N,S,E,W, wgs84
location = earth + wgs84.latlon(location_latitud * N,
location_longitud * E, elevation_m=0)
ts = load.timescale()
```

```
eph = load('de440s.bsp')
planets = load('de440s.bsp')
earth = planets['earth']
sun = planets['sun']
astro = location.at(ts.utc(year, month, day,
hour-timezone,minute)).observe(sun)
sun_app = astro.apparent()
sun_alt, sun_az, = sun_app.altaz()
```

Explanation of Each Line

`from skyfield.api import load:`

- Imports the load function from the skyfield.api module.
- The load function is used to fetch data files (e.g., timescales, ephemeris files) necessary for astronomical computations.

`from skyfield.api import N, S, E, W, wgs84:`

- Imports the constants N, S, E, W (representing the cardinal directions: North, South, East, and West) and the wgs84 object from the skyfield.api module.
- wgs84 is a geodetic model used for Earth-related computations, such as converting latitudes and longitudes to 3D coordinates.

`ts = load.timescale():`

- Loads a timescale object, which is used to work with time in Skyfield.
- The timescale object provides methods to define and manipulate time (e.g., UTC, TT).

`eph = load('de421.bsp'):`

- Loads the JPL (Jet Propulsion Laboratory) **DE421 ephemeris file**. This file contains precise positions and velocities for celestial bodies in the solar system.
- 'de421.bsp' is a binary file that Skyfield uses to compute positions of planets and other bodies.

`planets = load('de421.bsp'):`

- This line is redundant (repeats the loading of 'de421.bsp') but essentially assigns the same ephemeris data to a new variable planet.
- It's commonly done to improve code readability (e.g., distinguishing between different use cases for the same data file).

`earth = planets['earth']:`

- Extracts the Earth object from the ephemeris data. This object allows calculations involving Earth's position, including observer locations.

`sun = planets['sun']:`

- Extracts the sun object from the ephemeris data. This object can be used to calculate the sun's position relative to Earth or other bodies.

location = earth + wgs84.latlon(location_latitud * N, location_longitud * E, elevation_m=0)

`Earth:`

- Represents the Earth in the solar system as defined by the loaded ephemeris (de421.bsp).
- This object is used as the base reference for observer locations.

```
wgs84.latlon(location_latitud * N, location_longitud * E,
elevation_m=0):
```

- A geodetic model representing Earth's shape (latitude, longitude, elevation).
- Converts the geographical coordinates into a 3D position in space relative to Earth's center.

`location_latitud * N:`

- Multiplies the latitude value (location_latitud) by N (North), ensuring it's interpreted as a northern hemisphere coordinate. If in the southern hemisphere, you'd use S instead.

`location_longitud * E:`

- Multiplies the longitude value (location_longitud) by E (East). If in the western hemisphere, you'd use W.

`elevation_m=0:`

- Specifies the elevation (in meters) of the location. Here, it is set to 0, which corresponds to sea level.

`earth + ...:`

- Combines the Earth's position in the solar system with the observer's location on Earth. The resulting location object represents a specific point on Earth as it moves through space.

```
astro = location.at(ts.utc(year, month, day, hour,minu
te)).observe(sun)
```

`location.at(ts.utc(year, month, day, hour, minute)):`

- Computes the position of the specified location on Earth at the given **UTC time**.
- ts.utc(year, month, day, hour, minute): Defines the time in Coordinated Universal Time (UTC).
- This step determines where the Earth (and hence the observer) is in space at that moment.

`.observe(sun):`

- Calculates the **apparent position** of the sun as seen from the location at the specified time.
- Considers the relative positions of the observer, Earth, and sun.

```
sun_app = astro.apparent()
```

`astro.apparent():`

- Converts the **geometric position** of the sun (as computed by astro) into its **apparent position** by:

  1. Accounting for **light-time delay**: The sun's observed position includes the time it takes for light to travel from the sun to the observer.
  2. Including **aberration of light**: Adjusts for the motion of the Earth while observing the sun.

`sun_app:`

- This variable now contains the sun's **apparent position** as seen by the observer, accounting for these effects.

```
sun_alt, sun_az, sun_distance = sun_app.altaz()
```

Explanation of the Outputs:

`sun_alt (Altitude):`

- The angular height of the sun above or below the horizon (in degrees).
- Positive values: The sun is above the horizon.
- Zero: The sun is on the horizon (sunrise or sunset).
- Negative values: The sun is below the horizon (nighttime).

`sun_az (Azimuth):`

- The compass direction to the sun (in degrees).
- Measured clockwise from North:
  - 0° = North
  - 90° = East
  - 180° = South
  - 270° = West

`sun_distance (Distance):`

The distance to the sun from the observer's location, measured in Astronomical Units (AU).

1 AU ≈ 149.6 million kilometers, which is the average Earth-sun distance.

Exercise 2: Calculate the Solar Azimuth at the Latitude of 39.12 North, and Longitude of 80.11 East, in 6+ Timezone, on 13 April 2024, during 15:34

First, determine the sun azimuth. The initial step is to input the required variable.

```
import math
from skyfield.api import load, N, E, wgs84 # Skyfield
for astronomical calculations
import calendar # For handling dates and times

lat_location = 39.12
long_location = 80.11
timezone = 6
year = 2025
month = 4
day = 13
ele = 100
```

The idea is that the user can use the position of the sun to determine the direction of the Qibla without requiring timing of the Rashdul Qibla. We can use any time of the day when the sun is located above the horizon. For this case, we use the position of sun azimuth at 15:34. To input the variable for hour and minute,

```
hour = 15
minute = 34
```

And then, to determine the sun azimuth

```
# Load planetary ephemeris data (precise astronomical
positions)
eph = load('de440s.bsp')
planets = load('de440s.bsp')

# Get Earth and Sun objects from the ephemeris
earth = planets['earth']
sun = planets['sun']

location = earth + wgs84.latlon(lat_location, long_
location, elevation_m=0)

# Create timescale object and set observation time
ts = load.timescale()
t0 = ts.utc(year, month, day)
t1 = ts.utc(year, month, day + 1)
```

```
sun_astro = location.at(ts.utc(year, month, day,
hour-timezone, minute)).observe(sun)
sun_app = sun_astro.apparent()
sun_alt, sun_az, distance = sun_app.altaz()
print(sun_az.degrees)
```

From the sun azimuth, we can visualize the sun azimuth using the polar plot.

```
import matplotlib.pyplot as plt
import numpy as np

# Create polar plot (compass-style visualization)
fig, ax = plt.subplots(
    subplot_kw={'projection': 'polar'}, # Polar
    coordinate system
    figsize=(8, 8)                    # 8x8 inch figure
)

# Configure polar plot:
ax.set_theta_direction(-1)          # Clockwise rotation
                                     (standard for compass
                                     bearings)
ax.set_theta_zero_location('N')     # 0° at top (North)

# Convert sun direction from degrees to radians for
plotting
sun_direction_rad = np.deg2rad(sun_az.degrees)

# Plot a line from center (0,0) to edge (1) in sun's
direction
ax.plot(
    [0, sun_direction_rad],         # Angle in radians
    [0, 1],                     # Distance from center
    label=f'Sun Az: {sun_az.degrees:.2f}°',     # Legend
                                                  label
    color='orange',              # Color of line
    linewidth=2                  # Line thickness
)

# Display the plot
plt.show()
```

This results as shown in Figure 9.4.

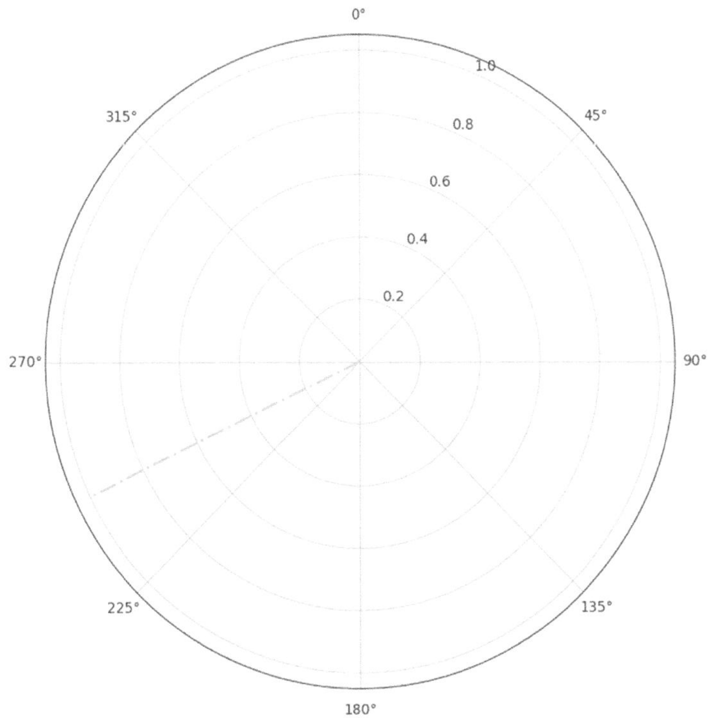

FIGURE 9.4 Polar plot with sun azimuth direction.

SUN AZIMUTH AND QIBLA DIRECTION VISUALIZATION ON POLAR PLOT

Now, let's combine polar plot for Qibla direction and sun azimuth.

Exercise 3: Visualize Qibla Direction and the Solar Azimuth on 13 April 2024, during 15:34, Using Location Latitude of 39.12 North, and Longitude of 80.11 East

The sun azimuth and Qibla direction visualization can be combined when both visualization codes are run on the same command console. First, run the Qibla direction calculation code.

```
φ_Location = 39.12
λ_Location = 80.11
φ_Kaabah = 21.4225
λ_Kaabah = 39.8262
Difference_Longitude = abs(λ_Location-λ_Kaabah)

#Calculation of Qibla Direction
import math

A = math.sin(math.radians(abs(Difference_Longitude)))
B = math.cos(math.radians(φ_Location))*math.tan(math.r
    adians(φ_Kaabah))
C = math.sin(math.radians(φ_Location)) * math.cos(m
    ath.radians(Difference_Longitude))
D = A/(B-C)
θ = math.degrees(math.atan(D))

#Determine the Azimuth of the Qibla
if Difference_Longitude > 180:
    delta_λ = 360 - Difference_Longitude
else:
    delta_λ = Difference_Longitude

if θ > 0:
    if λ_Location > λ_Kaabah:
        quadrant = "UB" # Utara Barat
    elif λ_Location <= λ_Kaabah:
        quadrant = "UT" # Utara Timur
    elif λ_Location < 0:
        if c >= 180:
            quadrant = "UB"
        else:
            quadrant = "UT"
elif θ < 0:
    if λ_Location > λ_Kaabah:
        quadrant = "SB" # Selatan Barat
    elif λ_Location <= λ_Kaabah:
        quadrant = "ST" # Selatan Timur
    elif λ_Location < 0:
        if c >= 180:
            quadrant = "SB"
        else:
            quadrant = "ST"

if quadrant == "UB":
    azimuth_kiblat = 360 - θ
```

```
elif quadrant == "SB":
    azimuth_kiblat = 180 - θ
elif quadrant == "UT":
    azimuth_kiblat = θ
elif quadrant == "ST":
    azimuth_kiblat = 180 + θ

# To Convert in Degree Form

degrees = int(azimuth_kiblat)
decimal_part = azimuth_kiblat - degrees
minutes_total = decimal_part * 60
minutes = int(minutes_total)
seconds = round((minutes_total - minutes) * 60)
print(f'The azimuth of the Qibla for Location with
coordinate {φ_Location} Latitude, {λ_Location}
Longitude, is {degrees}° {minutes}' {seconds}"')
```

Then the Sun Azimuth code

```
# Import required libraries
import math
from skyfield.api import load, N, E, wgs84 # Skyfield
for astronomical calculations
import calendar # For handling dates and times

lat_location = 39.12
long_location = 80.11
timezone = 6
year = 2025
month = 4
day = 13
ele = 100

hour = 15
minute = 34

# Load planetary ephemeris data (precise astronomical
positions)
eph = load('de440s.bsp')
planets = load('de440s.bsp')

# Get Earth and Sun objects from the ephemeris
earth = planets['earth']
sun = planets['sun']
```

```
location = earth + wgs84.latlon(lat_location, long_
location, elevation_m=0)

# Create timescale object and set observation time
ts = load.timescale()
t0 = ts.utc(year, month, day)
t1 = ts.utc(year, month, day + 1)

sun_astro = location.at(ts.utc(year, month, day,
hour-timezone, minute)).observe(sun)
sun_app = sun_astro.apparent()
sun_alt, sun_az, distance = sun_app.altaz()
print(sun_az.degrees)
```

Then run the combination of the visualization code. The combination code requires two important variables, sun azimuth as sun_az.degrees, and Qibla Direction as azimuth_kiblat.

```
import matplotlib.pyplot as plt
import numpy as np

fig, ax = plt.subplots(subplot_kw={'projection':
'polar'}, figsize=(8, 8))
ax.set_theta_direction(-1)
ax.set_theta_zero_location('N')

# Plot Qibla Direction (blue line)
qibla_direction_rad = np.deg2rad(azimuth_kiblat)
ax.plot([0, qibla_direction_rad], [0, 1], label=
f'Qibla: {azimuth_kiblat:.2f}°', color='blue',
linewidth=2)

# Plot Sun Azimuth (orange line)
sun_direction_rad = np.deg2rad(sun_az.degrees)
# Plot a line from center (0,0) to edge (1) in sun's
direction
ax.plot(
   [0, sun_direction_rad],       # Angle in radians
   [0, 1],                     # Distance from center
   label=f'Sun Az: {sun_az.degrees:.2f}°',      # Legend
                                                  label
   color='orange',              # Color of line
   linewidth=2                  # Line thickness
)
```

The resulting code generates a compass displaying both the Sun's azimuth and the Qibla direction. However, it currently lacks labels to help users interpret the visualization. To enhance its usability, the visual output should be updated with clear, descriptive labels.

```
import matplotlib.pyplot as plt
import numpy as np

fig, ax = plt.subplots(subplot_kw={'projection':
'polar'}, figsize=(8, 8))
ax.set_theta_direction(-1)
ax.set_theta_zero_location('N')

# Plot Qibla Direction (blue line)
qibla_direction_rad = np.deg2rad(azimuth_kiblat)
ax.plot([0, qibla_direction_rad], [0, 1], label=
f'Qibla: {azimuth_kiblat:.2f}°', color='blue',
linewidth=2)

# Plot Sun Azimuth (orange line)
sun_direction_rad = np.deg2rad(sun_az.degrees)
# Plot a line from center (0,0) to edge (1) in sun's
direction
ax.plot(
            [0, sun_direction_rad],[0, 1],label=f'Sun
            Az: {sun_az.degrees:.2f}°',
            color='orange',linewidth=2
)

plt.title(f'Visualization of Qibla and Sun Azimuth\
nLat: {lat_location}° | Long: {long_location}° \n{day}
{month} {year}, {hour}:{minute} Local Time\n', pad=20)
plt.legend(loc='upper right')
plt.show()
```

Figure 9.5 shows a polar plot; it's like a compass that shows two important directions: where the Qibla is (marked by the dashed line) and where the sun is at a particular time (shown with the dashed and dotted line). Right above the plot, you'll see the exact location it's based on, with the latitude and longitude, as well as the local date and time. That information is important because the direction of both the Qibla and the sun changes depending on where you are and what time it is.

Now, if you look around the circle, you'll notice numbers like 0°, 90°, 180°, and 270°. These represent directions: 0° is North, 90° is East, 180° is

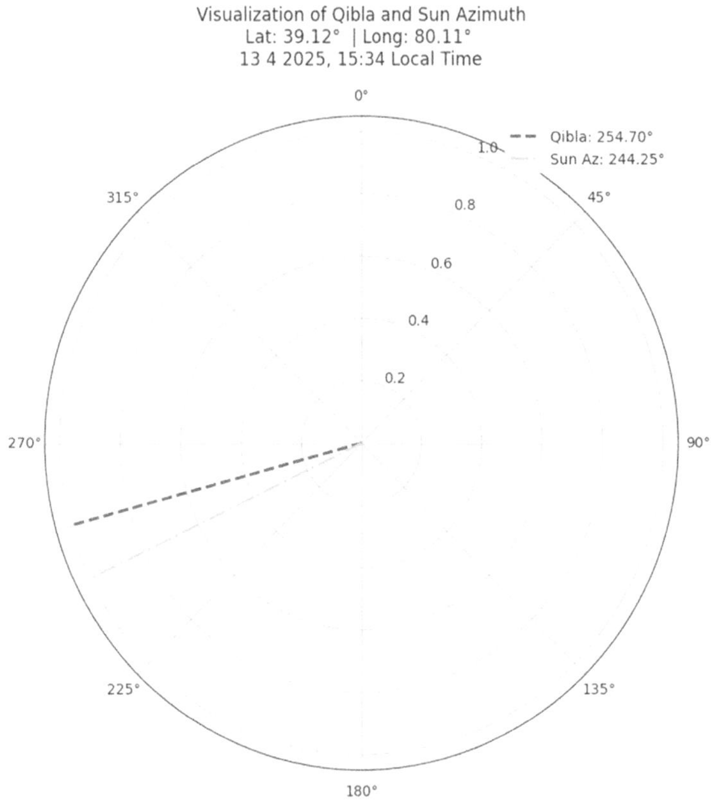

FIGURE 9.5 Polar plot of the sun azimuth and Qibla direction.

South, and 270° is West. So, imagine this as if you're standing inside a compass, and each degree tells you which way you're facing.

To use this, start by focusing on the blue line; that's the Qibla direction for your location. It tells you the exact angle you need to face to pray toward the Kaaba in Makkah. If you have a compass or a compass app on your phone, just rotate yourself until you're aligned with that degree. That's your Qibla.

Now here's where it gets even more helpful, notice the orange line? That shows where the sun is at the given time. If you're outdoors and the sun is visible, you can use its position to help you find the Qibla without needing a compass. One simple way is to use something like a water bottle or any object that stands upright. Place it on a flat surface and observe where its shadow falls. The shadow will point directly away from the sun's direction.

Now, look at the diagram again and find the orange line; that's the sun's azimuth. Then compare it to the blue line, which shows the Qibla direction. The angle between these two lines tells you how much you need to rotate from the sun's position to face the Qibla. For example, if the Qibla is to the left of the sun's position on the plot, you would turn your body that same amount to the left from the shadow's direction. This way, the shadow becomes your guide, and with just a bottle and this plot, you can figure out the Qibla direction even without technology. Pretty handy, right?

So, in short, this plot helps you see both where the Qibla is and where the sun is at a specific time, which can be useful, especially when you don't have a digital compass handy but can see the sun.

Exercise 4

Visualize the Qibla direction and the solar azimuth on August 17, 2025, at 14:40 local time, for a location with latitude 36.74° South and longitude 71.06° West (near Chillán, Chile). The location has an elevation of 150 meters above sea level, and the local timezone is UTC−4 (Chile Standard Time). Use this information to generate a polar plot showing both the direction of the Qibla and the azimuth of the sun at that moment.

Exercise 5

Visualize the Qibla direction and the solar azimuth on August 31, 2025, at 09:23 local time, for a location with latitude 48.85° North and longitude 2.35° East (Paris, France). The location has an elevation of 35 meters above sea level, and the local timezone is UTC+2 (Central European Summer Time). Generate a polar plot to visualize both the Qibla direction and the position of the sun at that specific time.

Exercise 6

Visualize the Qibla direction and the solar azimuth on October 10, 2025, at 16:10 local time, for a location with latitude 1.29° North and longitude 103.85° East (Singapore). The location has an elevation of 15 meters, and the local timezone is UTC+8 (Singapore Standard Time). Generate a plot that shows the sun's azimuth alongside the Qibla direction.

Exercise 7

Visualize the Qibla direction and the solar azimuth on December 5, 2025, at 07:00 local time, for a location with latitude 40.71° North and longitude –74.01° West (New York City, USA). The location has an elevation of 10 meters above sea level, and the local timezone is UTC–5 (Eastern Standard Time). Create a visualization to compare the sun's position with the Qibla direction during sunrise hours.

Sun Position during Prayer Times Visualization 10

VISUALIZING THE SUN POSITION

Prayer times are determined by the sun's position relative to the observer. Each prayer time corresponds to a specific solar event:

1. Zuhur begins when the sun reaches its highest point (zenith) in the sky, directly above the observer's location (solar noon).
2. Asr starts when the length of an object's shadow equals its actual height (or twice its height, depending on the school of thought).
3. Maghrib begins at sunset, when the sun completely disappears below the horizon.
4. Isha and Fajr depend on atmospheric twilight caused by the sun's diffraction below the horizon.
   - Isha begins when the sky is fully dark (astronomical twilight ends).
   - Fajr begins at dawn when the first light appears (astronomical twilight begins).
5. Sunrise marks the end of Fajr and the beginning of daytime.

This visualization helps illustrate these key solar positions, making it easier to understand the astronomical basis of Islamic prayer times. Before diving into the code, let's first understand the objective and the expected output:

DOI: 10.1201/9781003649120-10

A. Objective

We want to create a simple visualization that shows:

1. The position of an observer (stick figure) at a certain location.
2. The position of the sun in the sky at a specific altitude angle.
3. The horizon line as a reference (0°).

B. Expected Output

A plot containing the following elements:

1. Green line: Horizon (ground level).
2. Simple human figure: Representation of the observer.
3. Orange circle: The sun at a certain elevation.
4. Dashed line: Line of sight from the observer to the sun.

To visualize the position of the sun based on the observer, the step is as follows:

1. Import Libraries: Use matplotlib for plotting and numpy for mathematical calculations.
2. Setup Positions:
    - Observer coordinates (observer_x, observer_y).
    - Sun's altitude angle (altitude_angle).
3. Calculate Sun Position Based on the Angle (sun_x, sun_y).
4. Plot the Visualization:
    - Draw the horizon as a straight line.
    - Represent the sun as an orange dot.
    - Draw a stick figure (if an image is available).

Here is the Complete Code with Inline Explanations, but first, install related libraries.

Import Library

```
import matplotlib.pyplot as plt
import numpy as np
from PIL import Image
```

```
import requests
import io # Import io to handle the image data in
memory
```

Set Observer and Sun Positions

```
observer_x, observer_y = 5, 0 # Observer's position on
the horizon
altitude_angle = 40 # Angle in degrees (negative for
below horizon)
distance_to_sun = 3 # Arbitrary horizontal distance to
the Sun
```

Calculate the Sun's Position Based on the Angle

```
sun_x = observer_x - distance_to_sun # Place the Sun
to the left of the observer
sun_y = observer_y + np.tan(np.radians(altitude_
angle)) * distance_to_sun # Calculate vertical
position
```

Plot the Visualization

```
fig, ax = plt.subplots(figsize=(10, 6))

# Draw the horizon (straight line)
ax.plot([0, 10], [0, 0], color="green", linewidth=2,
label="Horizon")

# Add stick figure to the plot using direct download
link from Google Drive URL
image_url = "https://drive.google.com/uc?export
=download&id=1T7pLZNW6dF9PKxdOZ9UXZDbA84teRSdx"

# Download the image content from the URL
response = requests.get(image_url)
response.raise_for_status() # Raise an exception for
bad status codes
```

```
# Open the image from the downloaded content
stick_figure = Image.open(io.BytesIO(response.
content))

# Setup position of stick figure (on the horizon)
stick_x, stick_y = 5, 0
ax.imshow(stick_figure, extent=(stick_x - 0.3, stick_x
+ 0.3, stick_y, stick_y + 1))

# Add the Sun as an orange dot
ax.plot(sun_x, sun_y, marker="o", color="orange",
markersize=10, label="Sun")

# Draw the line of sight (dashed line)
ax.plot([observer_x, sun_x], [observer_y + 0.8,
sun_y], color="black", linestyle="--", linewidth=1,
label="Line of Sight")

# Add altitude scale
ax.axhline(0, color="black", linestyle="-",
linewidth=1)
ax.text(-0.5, 0, "Horizon (0°)", va="center",
ha="right", fontsize=10, color="green")
ax.text(-0.5, sun_y, f"Sun Position ({altitude_
angle}°)", va="center", ha="right", fontsize=10,
color="orange")

# Adjust the plot
ax.set_xlim(0, 10)
ax.set_ylim(-1.5, 5)

# Add labels and legend
ax.axis("off")
ax.legend(loc="upper right")
ax.set_title("Sun's Altitude Visualization",
fontsize=14)

plt.show()
```

The result of the code implementation above is shown in Figure 10.1.

This visualization demonstrates how solar altitude angles (like those used to determine prayer times) can be represented geometrically from an observer's perspective.

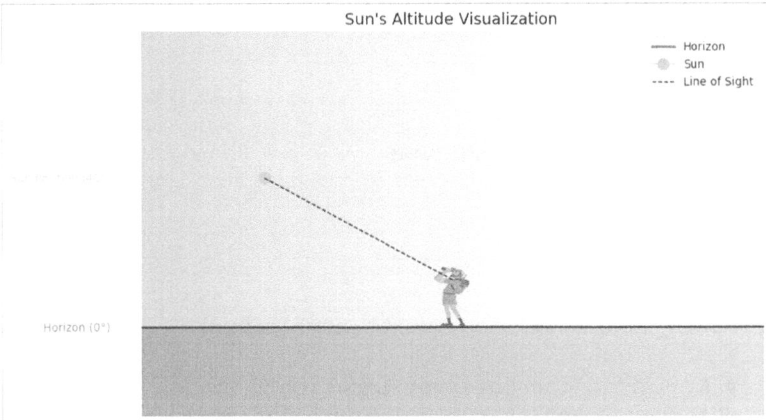

FIGURE 10.1 Visualization of sun position in respect to an observer.

Exercise 1: Calculating Sun's Altitude Angles for Prayer Times

Use Python's skyfield library to calculate the sun's altitude angle at specific prayer times (e.g., Asr) for given coordinates and dates.

For example, if asar prayer time on 13 November 2024, lat 3, long 101, timezone 8 is 4:32, the determination of the sun's altitude angle is:

Skyfield Installation

```
!pip install skyfield

# 2. Library Imports
from skyfield.api import load, wgs84
from skyfield.almanac import find_transits,find_
settings, find_risings
from datetime import datetime, timedelta
import math

# 3. Load Astronomical Data
ts = load.timescale()
eph = load('de440s.bsp')
sun = eph['Sun']
```

```
# 4. Define Observer's Location
latitude=3
longitude = 101
timezone = 8

# 5. Define Date and Time
day = 13
month = 11
year = 2024
hour = 16
minute = 32
month_name = calendar.month_name[month]

# 6. Initialize Observer Location
observer = eph['Earth'] + wgs84.latlon(latitude,
longitude)

# 7. Calculate Sun's Apparent Position
sun_astro = observer.at(ts.utc(year, month, day, hour,
minute)).observe(sun)
sun_app = sun_astro.apparent()
sun_alt, sun_az, distance = sun_app.altaz()

# 8. Output the Sun's Altitude
print(f'the Altitude of Sun at Asr prayer time on
{day} {month_name} {year}, at coordinate lat:
{latitude}, long: {longitude}, tz: {timezone}, at
{hour}:{minute} Local Time is {sun_alt.degrees}')
```

This Python code calculates the **altitude of the Sun** (in degrees) at a specific location, date, and time using the **Skyfield library**. Here's how the code works:

Library Imports

- **skyfield.api**: Provides tools to load astronomical data and calculate celestial positions.
- **skyfield.almanac**: Contains functions for calculating astronomical events, though not directly used here.
- **datetime, timedelta**: Handles date and time manipulation.
- **math**: Not used in this snippet but available for rounding operations if needed.

Load Astronomical Data

```
ts = load.timescale()
eph = load('de440s.bsp')
sun = eph['Sun']
```

- **ts = load.timescale()**: Initializes a timescale object for handling time-related calculations.
- **eph = load('de440s.bsp')**: Loads the DE440s ephemeris file containing accurate positional data for celestial objects.
- **sun = eph['Sun']**: Loads positional data for the Sun.

Define Observer's Location

```
latitude = 3
longitude = 101
timezone = 8
```

- **latitude and longitude**: Specify the geographic coordinates of the observer. These coordinates correspond to a location in Malaysia.
- **timezone**: The local timezone offset from UTC (Malaysia = +8).

Define Date and Time

```
day = 13
month = 6
year = 2024
hour = 16
minute = 32
```

- Specifies the exact **date** (June 13, 2024) and **time** (16:32 in local time) for which the sun's altitude will be calculated.

Initialize Observer Location

```
observer = eph['Earth'] + wgs84.latlon(latitude,
longitude)
```

- Combines the Earth's position with the observer's location using the **WGS84 geodetic system**.

Calculate Sun's Apparent Position

```
sun_astro = observer.at(ts.utc(year, month, day, hour
- timezone, minute)).observe(sun)
sun_app = sun_astro.apparent()
```

- **ts.utc(year, month, day, hour – timezone, minute)**: Converts the local time to UTC by subtracting the timezone offset (16:32 local time → 08:32 UTC).
- **observer.at()**: Specifies the observer's position at the given UTC time.
- **observe(sun)**: Calculates the Sun's relative position in the sky from the observer's location.
- **apparent()**: Adjusts for atmospheric effects (e.g., refraction) to provide the apparent position of the Sun.

Calculate Sun's Altitude

```
sun_alt, sun_az, distance = sun_app.altaz()
```

- **altaz()**: Computes the Sun's **altitude**, **azimuth**, and **distance** from the observer.
    - **sun_alt**: The altitude of the Sun in degrees (angle above the horizon).
    - **sun_az**: The azimuth of the Sun in degrees (angle measured clockwise from true north).
    - **distance**: The distance between the observer and the Sun (not used here).

Output the Sun's Altitude

```
print(f'the Altitude of Sun during Asr prayer time on
{day} {month_name} {year}, at coordinate lat:
{latitude}, long: {longitude}, tz: {timezone}, at
{hour}:{minute} Local Time is {sun_alt.degrees}')
```

- **sun_alt.degrees**: Converts the altitude from radians (default Skyfield unit) to degrees and prints it.
- The value represents the sun's angular elevation above the horizon:

```
the Altitude of sun during Asr prayer time on 13
November 2024, at coordinate lat: 3, long: 101, tz: 8,
at 16:32 Local Time is 33.77219002658052
```

VISUALIZATION OF SUN POSITION DURING ZUHUR PRAYER TIME

Exercise 2: Visualizing the Sun's Altitude at Zuhur Prayer Time for the Given Location and Date

Create a visualization that shows the sun's position in the sky at Zuhur prayer time for the given coordinates (latitude: 39.9°N, longitude: 116.4°E, time zone: UTC+8) on December 19, 2024, with elevation of 100 m.

First, determine the time of Zuhur using the given location.

```
#Import Necessary Function
from skyfield.api import load
from skyfield.api import N, S, E, W, wgs84
from skyfield import almanac

#Load Ephemeris Data and Planet Objects
ts = load.timescale()
eph = load('de440s.bsp')
planets = load('de440s.bsp')
earth = planets['earth']
sun = planets['sun']

# Variable Input

lat_location = 39.9
long_location = 116.4
timezone = 8
day = 19
month = 12
year = 2024
ele = 100
```

```
#input into
location = earth + wgs84.latlon(lat_location, long_
location, elevation_m=0)

#Range of Data

t0 = ts.utc(year, month, day)
t1 = ts.utc(year, month, day + 1)

# Time of Solar Transit
t = almanac.find_transits(location, sun, t0, t1)
hour_solar_transit = t.utc.hour
minutes_solar_transit = t.utc.minute
second_solar_transit = t.utc.second

zuhur_time = hour_solar_transit + (minutes_solar_
transit / 60) + (second_solar_transit / 3600 ) +
timezone + 0.017778

zuhur_time = float(zuhur_time)
degrees = int(zuhur_time )
decimal_part = zuhur_time - degrees
minutes_total = decimal_part * 60
minutes = int(minutes_total)

seconds = round((minutes_total - minutes) * 60)
sun_astro = location.at(ts.utc(year, month, day,
hour_solar_transit, minutes_solar_transit, second_
solar_transit)).observe(sun)
sun_alt, _, _ = sun_astro.apparent().altaz()
# Check if the sun is above the horizon at zuhur time
if sun_alt.degrees <= 0:
    zuhur = "Zuhur Does Not Occur"
else:
    zuhur = f"Zuhur Occurs at {degrees}° {minutes}'
    {seconds}""

print(zuhur)
Zuhur Occurs at 12° 12' 36"
```

The calculated Zuhur prayer time for the specified location is 12:12:36. This result is derived from the values stored in the variables: degrees = 12, minutes = 12, and seconds = 36. These values represent the time when the sun has just passed its daily zenith. It is important to note that this calculation assumes conversion based on the local time (LTC) using the appropriate time zone offset.

However, if we intend to use these values for further astronomical calculations, such as determining the solar position, we must ensure that the computation is based on UTC (Universal Coordinated Time). This is crucial because astronomical algorithms, such as those used to determine solar altitude or azimuth, typically rely on standardized time references like UTC to maintain accuracy and consistency across different locations and dates.

From the given time, determine the position of the sun.

```
# Local time values
h = degrees
m = minutes
s = seconds

# Adjust local time to UTC
h_utc = h - timezone

# Compute observation time using Skyfield
sun_astro = location.at(ts.utc(year, month, day,
h_utc, m, s)).observe(sun)
sun_alt, _, _ = sun_astro.apparent().altaz()

print(sun_alt)
26deg 40' 41.6"
```

Now we get the position of the sun is 26deg 40' 41.6" of solar altitude. We can use this value to visualize the position of the sun. Visualization of the sun can be used using matplotlib. The code could be a little overwhelming, but the important thing is the variable.

```
# 1. Import Library
import matplotlib.pyplot as plt
import numpy as np
from PIL import Image
import requests
import io # Import io to handle the image data in
memory

# 2. Set Observer and Sun Positions
observer_x, observer_y = 5, 0 # Observer's position on
the horizon
altitude_angle = sun_alt.degrees[0] # Angle in degrees
(negative for below horizon)
distance_to_sun = 3 # Arbitrary horizontal distance to
the Sun
```

```
# 3. Calculate the Sun's position based on the angle
sun_x = observer_x - distance_to_sun # Place the Sun
to the left of the observer
sun_y = observer_y + np.tan(np.radians(altitude_
angle)) * distance_to_sun # Calculate vertical
position

# 4. Plot the Visualization
fig, ax = plt.subplots(figsize=(10, 6))

# Draw the horizon (straight line)
ax.plot([0, 10], [0, 0], color="green", linewidth=2,
label="Horizon")

# Add stick figure to the plot using direct download
link from Google Drive URL
image_url = "https://drive.google.com/uc?export
=download&id=1T7pLZNW6dF9PKxdOZ9UXZDbA84teRSdx"
# Download the image content from the URL
response = requests.get(image_url)
response.raise_for_status() # Raise an exception for
bad status codes
# Open the image from the downloaded content
stick_figure = Image.open(io.BytesIO(response.
content))

stick_x, stick_y = 5, 0 # Position of stick figure (on
the horizon)
ax.imshow(stick_figure, extent=(stick_x - 0.3, stick_x
+ 0.3, stick_y, stick_y + 1))

# Add the Sun as an orange dot
ax.plot(sun_x, sun_y, marker="o", color="orange",
markersize=10, label="Sun")

# Add sky gradient (blue)
sky_gradient = np.linspace(1, 0, 256).reshape(1, -1)
sky_gradient = np.vstack((sky_gradient, sky_gradient))
ax.imshow(sky_gradient, extent=[0, 10, -1.5, 5],
cmap='Blues', alpha=0.3, aspect='auto')

# Add ground (green)
ground = plt.Rectangle((0, -1.5), 10, 1.5,
color='darkgreen', alpha=0.3)
ax.add_patch(ground)
```

```
# Draw the line of sight (dashed line)
ax.plot([observer_x, sun_x], [observer_y + 0.8,
sun_y], color="black", linestyle="--", linewidth=1,
label="Line of Sight")

# Add altitude scale
ax.axhline(0, color="black", linestyle="-",
linewidth=1)
ax.text(-0.5, 0, "Horizon (0°)", va="center",
ha="right", fontsize=10, color="green")
ax.text(-0.5, sun_y, f"Sun Position ({altitude_
angle:.4f}°)", va="center", ha="right", fontsize=10,
color="orange")

# Adjust the plot
ax.set_xlim(0, 10)
ax.set_ylim(-1.5, 5)

# Add labels and legend
ax.axis("off")
ax.legend(loc="upper right")
ax.set_title(f"Sun's Altitude Visualization at Zuhur
Prayer Time\n Lat: {lat_location}° N Long: {long_
location}° E TZ: {timezone}\n {day} {month_name}
{year} {degrees}: {minutes}: {seconds}", fontsize=14)

plt.show()
```

In the code above, the most important variable is sun_alt.degrees, which represents the sun's altitude in degrees at the specified moment. This value is critical for determining whether the sun meets the required condition for a given prayer time. The remaining code functions primarily as a template, providing structure for observation and visualization, and does not directly affect the outcome of the sun's altitude calculation. Additionally, when visualizing different prayer times (e.g., Subh, Zuhur, Asar, etc.), ensure that the graph title is updated accordingly by modifying the line:

```
ax.set_title(f"Sun's Altitude Visualization at Zuhur
Prayer Time\n Lat: {lat_location}° N Long: {long_
location}° E TZ: {timezone}\n {day} {month_name} {year}
{degrees}: {minutes}: {seconds}", fontsize=14)
```

to reflect the **specific prayer time** being analyzed. The result of the coding visualization is shown in Figure 10.2.

FIGURE 10.2 Visualization of sun position during Zuhr prayer time.

VISUALIZATION OF SUN POSITION DURING ASAR PRAYER TIME

Exercise 3: Visualizing the Sun's Altitude at Asar Prayer Time for the Given Location and Date

Create a visualization that shows the sun's position in the sky at Asar prayer time for the given coordinates (latitude: 39.9°N, longitude: 116.4°E, time zone: UTC+8) on December 19, 2024. In exercise 3, the given data only includes location coordinates and dates. Therefore, we need to calculate the sun's altitude at the time of Asar as explained in Chapter 7. After that, we can then create a visualization of the sun's altitude at the time of Asar. First, calculate the time of Asar prayer time

```
#Calculate Asr Prayer Time and Sun's Altitude
# 1. Import Necessary Function
from skyfield.api import load
from skyfield.api import N, S, E, W, wgs84
from skyfield import almanac
import math
import calendar
```

```
# 2. Define Observer's Location
lat_location = 39.9
long_location = 116.4
timezone = 8

# 3. Define Date
day = 19
month = 12
year = 2024
month_name = calendar.month_name[month]

# 4. Load Astronomical Data
ts = load.timescale()
eph = load('de440s.bsp')
planets = load('de440s.bsp')
earth = planets['earth']
sun = planets['sun']

# 6. Initialize Observer Location
location = earth + wgs84.latlon(lat_location, long_
location, elevation_m=0)

# 7. Setup range of calculate date for calculating
solar transit
t0 = ts.utc(year, month, day)
t1 = ts.utc(year, month, day + 1)

# 8. Calculate the time of solar transit
t = almanac.find_transits(location, sun, t0, t1)

# 9. The position of sun altitude at the time of the
solar transit
h, m, s = t.utc.hour, t.utc.minute, t.utc.second
sun_astro = location.at(ts.utc(year, month, day, h,
m)).observe(sun)
sun_app = sun_astro.apparent()
sun_alt, sun_az, distance = sun_app.altaz()

# 10. Calculate the length of the sun shadow during
transit
sun_shadow_transit = 1/(math.tan(math.radians(sun_a
lt.degrees)))

# 11. Calculate the length of the sun shadow at Asr
Prayer Time
sun_shadow_asar = 1 + sun_shadow_transit
```

```
# 12. Loop the sun shadow does not pass the length of
the asar sun shadow
# Start with hour
# Start with hour

test = 1
while True:

    # Calculate the shadow length
    sun_astro = location.at(ts.utc(year, month, day, h,
    m, s)).observe(sun)
    sun_alt, _, _ = sun_astro.apparent().altaz() # Get
    the altitude of the sun
    sun_shadow= 1 / math.tan(math.radians(sun_alt.
    degrees))

    if sun_alt.degrees <= 0:
      break
    if test > 24:
      break

    if sun_shadow >= sun_shadow_asar:
      break # Exit the loop if the shadow length
      matches or exceeds the desired length
    h += 1

# Once the condition is met for hours, move to minutes
h_asar = h - 1
test=1
while True:
    # Calculate the shadow length
    sun_astro = location.at(ts.utc(year, month, day,
    h_asar, m, s)).observe(sun)
    sun_alt, _, _ = sun_astro.apparent().altaz() # Get
    the altitude of the sun
    sun_shadow = 1 / math.tan(math.radians(sun_alt.
    degrees))

    if sun_alt.degrees <= 0:
      break
    if test > 1440:
      break

    if sun_shadow >= sun_shadow_asar:
      break # Exit the loop if the shadow length
      matches or exceeds the desired length
    m += 1
```

```
    # Increment time in minutes
m_asar = m - 1
test = 1

# Once the condition is met for minutes, move to
seconds
while True:
    # Calculate the shadow length
    sun_astro = location.at(ts.utc(year, month, day,
    h_asar, m_asar, s)).observe(sun)
    sun_alt, _, _ = sun_astro.apparent().altaz() # Get
    the altitude of the sun
    sun_shadow = 1 / math.tan(math.radians(sun_alt.
    degrees))

    if sun_alt.degrees <= 0:
      break
    if test > 86400:
      break

    if sun_shadow >= sun_shadow_asar:
      break # Exit the loop if the shadow length
      matches or exceeds the desired length
    s += 1

    # Increment time in seconds

s_asar = s

asar_time = (h_asar + (m_asar) / 60 + s_asar / 3600) +
timezone
asar_time = float(asar_time)
degrees = int(asar_time)
decimal_part = asar_time - degrees
minutes_total = decimal_part * 60
minutes = int(minutes_total)
seconds = round((minutes_total - minutes) * 60)

if sun_alt.degrees <= 0 or test >86400:
     asar = "Asar Does Not Occur"
else:
     asar = f"Asar Occurs at {degrees}° {minutes}'
     {seconds}""

print(asar)
Asar Occurs at 14° 34' 7"
```

Second, sun's altitude visualization,

```
# 1. Import Library
import matplotlib.pyplot as plt
import numpy as np
from PIL import Image
import requests
import io # Import io to handle the image data in
memory

# 2. Set Observer and Sun Positions
observer_x, observer_y = 5, 0 # Observer's position on
the horizon
altitude_angle = sun_alt.degrees[0] # Angle in degrees
(negative for below horizon)
distance_to_sun = 3 # Arbitrary horizontal distance to
the Sun

# 3. Calculate the Sun's position based on the angle
sun_x = observer_x - distance_to_sun # Place the Sun
to the left of the observer
sun_y = observer_y + np.tan(np.radians(altitude_
angle)) * distance_to_sun # Calculate vertical
position

# 4. Plot the Visualization
fig, ax = plt.subplots(figsize=(10, 6))

# Draw the horizon (straight line)
ax.plot([0, 10], [0, 0], color="green", linewidth=2,
label="Horizon")

# Add stick figure to the plot using direct download
link from Google Drive URL
image_url = "https://drive.google.com/uc?export
=download&id=1T7pLZNW6dF9PKxdOZ9UXZDbA84teRSdx"
# Download the image content from the URL
response = requests.get(image_url)
response.raise_for_status() # Raise an exception for
bad status codes
# Open the image from the downloaded content
stick_figure = Image.open(io.BytesIO(response.
content))

stick_x, stick_y = 5, 0 # Position of stick figure (on
the horizon)
```

```
ax.imshow(stick_figure, extent=(stick_x - 0.3, stick_x
+ 0.3, stick_y, stick_y + 1))

# Add the Sun as an orange dot
ax.plot(sun_x, sun_y, marker="o", color="orange",
markersize=10, label="Sun")

# Add sky gradient (blue)
sky_gradient = np.linspace(1, 0, 256).reshape(1, -1)
sky_gradient = np.vstack((sky_gradient, sky_gradient))
ax.imshow(sky_gradient, extent=[0, 10, -1.5, 5],
cmap='Blues', alpha=0.3, aspect='auto')

# Add ground (green)
ground = plt.Rectangle((0, -1.5), 10, 1.5,
color='darkgreen', alpha=0.3)
ax.add_patch(ground)
# Draw the line of sight (dashed line)
ax.plot([observer_x, sun_x], [observer_y + 0.8,
sun_y], color="black", linestyle="--", linewidth=1,
label="Line of Sight")

# Add altitude scale
ax.axhline(0, color="black", linestyle="-",
linewidth=1)
ax.text(-0.5, 0, "Horizon (0°)", va="center",
ha="right", fontsize=10, color="green")
ax.text(-0.5, sun_y, f"Sun Position ({altitude_
angle:.4f}°)", va="center", ha="right", fontsize=10,
color="orange")

# Adjust the plot
ax.set_xlim(0, 10)
ax.set_ylim(-1.5, 5)

# Add labels and legend
ax.axis("off")
ax.legend(loc="upper right")
ax.set_title(f"Sun's Altitude Visualization at Asr
Prayer Time\n Lat: {lat_location}° N Long: {long_
location}° E TZ: {timezone}\n {day} {month_name}
{year} {degrees}: {minutes}: {seconds}", fontsize=14)

plt.show()
```

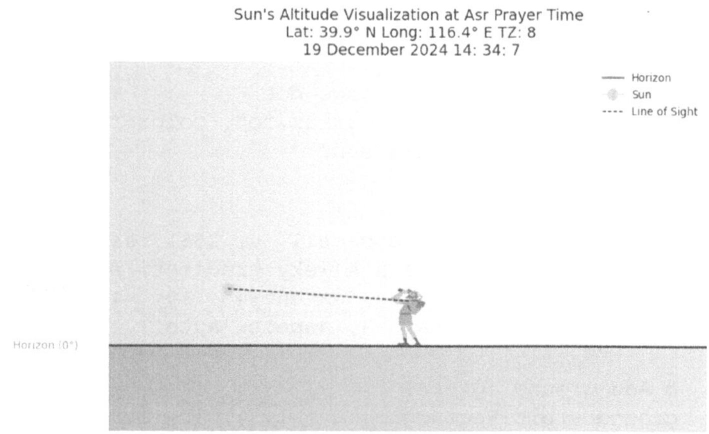

FIGURE 10.3 Visualization of sun position during Asar prayer time.

The result of the code implementation above is shown in Figure 10.3.

VISUALIZATION OF SUN POSITION DURING MAGHRIB PRAYER TIME

Exercise 4: Visualizing the Sun's Altitude at Maghrib Prayer Time for the Given Location and Date

Create a visualization that shows the sun's position in the sky at Maghrib prayer time for the given coordinates (latitude: 39.9°N, longitude: 116.4°E, time zone: UTC+8) on December 19, 2024, with elevation of 100 m. First, determine the maghrib prayer time.

```
#Import Necessary Function
from skyfield.api import load
from skyfield.api import N, S, E, W, wgs84
from skyfield import almanac

#Load Ephemeris Data and Planet Objects
ts = load.timescale()
```

```
eph = load('de440s.bsp')
planets = load('de440s.bsp')
earth = planets['earth']
sun = planets['sun']

# Variable Input
lat_location = 39.9
long_location = 116.4
timezone = 8
day = 19
month = 12
year = 2024
ele = 100

#input into
location = earth + wgs84.latlon(lat_location, long_
location, elevation_m=0)

#Range of Data

t0 = ts.utc(year, month, day)
t1 = ts.utc(year, month, day + 1)

from skyfield.units import Angle
from numpy import arccos
from skyfield.earthlib import refraction

altitude_m = ele
earth_radius_m = 6378136.6
side_over_hypotenuse = earth_radius_m / (earth_
radius_m + altitude_m)
h = Angle(radians=-arccos(side_over_hypotenuse))
solar_radius_degrees = 16 / 60
r = refraction(0.0, temperature_C=15.0,
pressure_mbar=1030.0)

t, y = almanac.find_settings(location, sun, t0, t1,
horizon_degrees=-r + h.degrees - solar_radius_degrees)
h, m, s = t.utc.hour, t.utc.minute, t.utc.second

maghrib_time = float(h + m / 60 + s / 3600 + timezone)
maghrib_time %= 24 # Ensure 24-hour clock format
maghrib_time = float(maghrib_time)
degrees = int(maghrib_time)
decimal_part = maghrib_time - degrees
```

```
minutes_total = decimal_part * 60
minutes = int(minutes_total)
seconds = round((minutes_total - minutes) * 60)

sun_astro = location.at(ts.utc(year, month, day, h, m,
s)).observe(sun)
sun_alt, _, _ = sun_astro.apparent().altaz()
if sun_alt.degrees >= 0:
    maghrib = "Maghrib Does Not Occur"
else:
    maghrib = f"Maghrib Occurs at {degrees}° {minutes}'
    {seconds}""

print(maghrib)
Maghrib Occurs at 16° 53' 43"
```

First, determine sun altitude.

```
# Local time values
h = degrees
m = minutes
s = seconds

# Adjust local time to UTC
h_utc = h - timezone

# Compute observation time using Skyfield
sun_astro = location.at(ts.utc(year, month, day,
h_utc, m, s)).observe(sun)
sun_alt, _, _ = sun_astro.apparent().altaz()

print(sun_alt)
-01deg 09' 42.5"
```

Then visualize the position of the sun

```
# 1. Import Library
import matplotlib.pyplot as plt
import numpy as np
from PIL import Image
import requests
import io # Import io to handle the image data in
memory
```

```
# 2. Set Observer and Sun Positions
observer_x, observer_y = 5, 0 # Observer's position on
the horizon
altitude_angle = sun_alt.degrees # Angle in degrees
(negative for below horizon)
distance_to_sun = 3 # Arbitrary horizontal distance to
the Sun

# 3. Calculate the Sun's position based on the angle
sun_x = observer_x - distance_to_sun # Place the Sun
to the left of the observer
sun_y = observer_y + np.tan(np.radians(altitude_
angle)) * distance_to_sun # Calculate vertical
position

# 4. Plot the Visualization
fig, ax = plt.subplots(figsize=(10, 6))

# Draw the horizon (straight line)
ax.plot([0, 10], [0, 0], color="green", linewidth=2,
label="Horizon")

# Add stick figure to the plot using direct download
link from Google Drive URL
image_url = "https://drive.google.com/uc?export
=download&id=1T7pLZNW6dF9PKxdOZ9UXZDbA84teRSdx"
# Download the image content from the URL
response = requests.get(image_url)
response.raise_for_status() # Raise an exception for
bad status codes
# Open the image from the downloaded content
stick_figure = Image.open(io.BytesIO(response.
content))

stick_x, stick_y = 5, 0 # Position of stick figure (on
the horizon)
ax.imshow(stick_figure, extent=(stick_x - 0.3, stick_x
+ 0.3, stick_y, stick_y + 1))

# Add the Sun as an orange dot
ax.plot(sun_x, sun_y, marker="o", color="orange",
markersize=10, label="Sun")

# Add sky gradient (blue)
sky_gradient = np.linspace(1, 0, 256).reshape(1, -1)
```

```
sky_gradient = np.vstack((sky_gradient, sky_gradient))
ax.imshow(sky_gradient, extent=[0, 10, -1.5, 5],
cmap='Blues', alpha=0.3, aspect='auto')

# Add ground (green)
ground = plt.Rectangle((0, -1.5), 10, 1.5,
color='darkgreen', alpha=0.3)
ax.add_patch(ground)

# Draw the line of sight (dashed line)
ax.plot([observer_x, sun_x], [observer_y + 0.8,
sun_y], color="black", linestyle="--", linewidth=1,
label="Line of Sight")

# Add altitude scale
ax.axhline(0, color="black", linestyle="-",
linewidth=1)
ax.text(-0.5, 0, "Horizon (0°)", va="center",
ha="right", fontsize=10, color="green")
ax.text(-0.5, sun_y-1, f"Sun Position ({altitude_
angle:.4f}°)", va="center", ha="right", fontsize=10,
color="orange")

# Adjust the plot
ax.set_xlim(0, 10)
ax.set_ylim(-1.5, 5)
# Add labels and legend
```

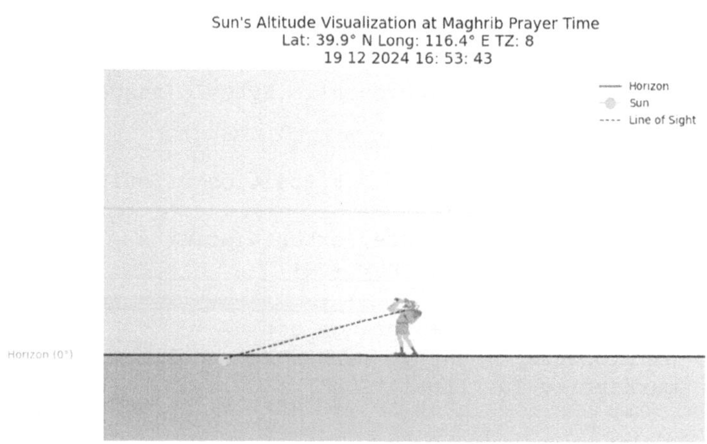

FIGURE 10.4 Visualization of sun position during Maghrib prayer time.

```
ax.axis("off")
ax.legend(loc="upper right")
ax.set_title(f"Sun's Altitude Visualization at Maghrib
Prayer Time\n Lat: {lat_location}° N Long: {long_
location}° E TZ: {timezone}\n {day} {month} {year}
{degrees}: {minutes}: {seconds}", fontsize=14)

plt.show()
```

The result of the code implementation above is shown in Figure 10.4.

VISUALIZATION OF SUN POSITION DURING ISYA' PRAYER TIME

Exercise 5: Visualizing the Sun's Altitude at Isya' Prayer Time for the Given Location and Date

Create a visualization that shows the sun's position in the sky at Isya' prayer time for the given coordinates (latitude: 39.9°N, longitude: 116.4°E, time zone: UTC+8) on December 19, 2024, for solar depression degree of 16. First, determine the Isya' prayer time

```
#Import Necessary Function
from skyfield.api import load
from skyfield.api import N, S, E, W, wgs84
from skyfield import almanac
import math

#Load Ephemeris Data and Planet Objects
ts = load.timescale()
eph = load('de440s.bsp')
planets = load('de440s.bsp')
earth = planets['earth']
sun = planets['sun']

# Variable Input

lat_location = 39.9
long_location = 116.4
```

```
timezone = 8
day = 19
month = 12
year = 2024
ele = 100

location = earth + wgs84.latlon(lat_location, long_
location, elevation_m=ele)
t0 = ts.utc(year, month, day)
t1 = ts.utc(year, month, day + 1)

from skyfield.units import Angle
from numpy import arccos
from skyfield.earthlib import refraction

altitude_m = ele
earth_radius_m = 6378136.6
side_over_hypotenuse = earth_radius_m / (earth_
radius_m + altitude_m)
h = Angle(radians=-arccos(side_over_hypotenuse))
solar_radius_degrees = 16 / 60
r = refraction(0.0, temperature_C=15.0,
pressure_mbar=1030.0)

t, y = almanac.find_settings(location, sun, t0, t1,
horizon_degrees=-r + h.degrees - solar_radius_degrees)
h, m, s = t.utc.hour, t.utc.minute, t.utc.second
#print(h,m,s)

m=1
s=1
h=h+1

sun_astro = location.at(ts.utc(year, month, day, h,
m)).observe(sun)
sun_app = sun_astro.apparent()
sun_alt, sun_az, distance = sun_app.altaz()
#print(sun_alt)

# Start with hour

isya_angle = 16
elevation_correction = 0.0293 * math.sqrt(ele)
isha_angle_actual = -isya_angle -elevation_correction
#print(isya_angle_corrected)
```

```
# Start with hour

test = 1
while True:

   # Calculate the Solar Altitude
   sun_astro = location.at(ts.utc(year, month, day,
   h+1, m, s)).observe(sun)
   sun_alt, _, _ = sun_astro.apparent().altaz() # Get
   the altitude of the sun

   if sun_alt.degrees >= 0:
     break
   if test > 24:
     break

   if sun_alt.degrees <= isha_angle_actual:
       break # Exit the loop if the solar altitude
       located below -18 degree
   h =h+ 1

# Once the condition is met for hours, move to minutes
h_isya = h - 1
test = 1
while True:
   # Calculate the Solar Altitude
   sun_astro = location.at(ts.utc(year, month, day,
   h_isya, m, s)).observe(sun)
   sun_alt, _, _ = sun_astro.apparent().altaz() # Get
   the altitude of the sun

   if sun_alt.degrees >= 0:
     break
   if test > 1440:
     break

   if sun_alt.degrees <= isha_angle_actual:
       break # Exit the loop if the solar altitude
       located below -18 degree
   m += 1

   # Increment time in minutes
m_isya = m - 1
```

```
# Once the condition is met for minutes, move to
seconds
test = 1
while True:
   # Calculate the Solar Altitude
   sun_astro = location.at(ts.utc(year, month, day,
   h_isya, m_isya, s)).observe(sun)
   sun_alt, _, _ = sun_astro.apparent().altaz() # Get
   the altitude of the sun

   if sun_alt.degrees >= 0:
     break
   if test > 86400:
     break

   if sun_alt.degrees <= isha_angle_actual:
       break # Exit the loop if the solar altitude
       located below -18 degree
   s += 1

   # Increment time in seconds

s_isya = s

isya_time = float(h_isyak + m_isyak / 60 + s_isyak /
3600 + timezone)
isya_time %= 24 # Ensure 24-hour clock format
isya_time = float(isya_time)
degrees = int(isya_time)
decimal_part = isya_time - degrees
minutes_total = decimal_part * 60
minutes = int(minutes_total)
seconds = round((minutes_total - minutes) * 60)

if sun_alt.degrees >= 0 or test >86400:
    isya = "Isya' Does Not Occur"
else:
    isya = f"Isya' Occurs at {degrees}° {minutes}'
    {seconds}""

print(isya)
Isya' Occurs at 18° 19' 59"
```

Determine the sun altitude during Isya' Prayer Time

```
# Local time values
h = degrees
m = minutes
s = seconds

# Adjust local time to UTC
h_utc = h - timezone

# Compute observation time using Skyfield
sun_astro = location.at(ts.utc(year, month, day,
h_utc, m, s)).observe(sun)
sun_alt, _, _ = sun_astro.apparent().altaz()

print(sun_alt)
-16deg 17′ 40.3″
```

Visualize the position of the sun during Isya' Prayer Time

```
# 1. Import Library
import matplotlib.pyplot as plt
import numpy as np
from PIL import Image
import requests
import io # Import io to handle the image data in
memory

# 2. Set Observer and Sun Positions
observer_x, observer_y = 5, 0 # Observer's position on
the horizon
altitude_angle = sun_alt.degrees # Angle in degrees
(negative for below horizon)
distance_to_sun = 3 # Arbitrary horizontal distance to
the Sun

# 3. Calculate the Sun's position based on the angle
sun_x = observer_x - distance_to_sun # Place the Sun
to the left of the observer
sun_y = observer_y + np.tan(np.radians(altitude_
angle)) * distance_to_sun # Calculate vertical
position
```

```
# 4. Plot the Visualization
fig, ax = plt.subplots(figsize=(10, 6))

# Draw the horizon (straight line)
ax.plot([0, 10], [0, 0], color="green", linewidth=2,
label="Horizon")

# Add stick figure to the plot using direct download
link from Google Drive URL
image_url = "https://drive.google.com/uc?export
=download&id=1T7pLZNW6dF9PKxdOZ9UXZDbA84teRSdx"
# Download the image content from the URL
response = requests.get(image_url)
response.raise_for_status() # Raise an exception for
bad status codes
# Open the image from the downloaded content
stick_figure = Image.open(io.BytesIO(response.
content))

stick_x, stick_y = 5, 0 # Position of stick figure (on
the horizon)
ax.imshow(stick_figure, extent=(stick_x - 0.3, stick_x
+ 0.3, stick_y, stick_y + 1))

# Add the Sun as an orange dot
ax.plot(sun_x, sun_y, marker="o", color="orange",
markersize=10, label="Sun")

# Add sky gradient (blue)
sky_gradient = np.linspace(1, 0, 256).reshape(1, -1)
sky_gradient = np.vstack((sky_gradient, sky_gradient))
ax.imshow(sky_gradient, extent=[0, 10, -1.5, 5],
cmap='Blues', alpha=0.3, aspect='auto')

# Add ground (green)
ground = plt.Rectangle((0, -1.5), 10, 1.5,
color='darkgreen', alpha=0.3)
ax.add_patch(ground)

# Draw the line of sight (dashed line)
ax.plot([observer_x, sun_x], [observer_y + 0.8,
sun_y], color="black", linestyle="--", linewidth=1,
label="Line of Sight")
```

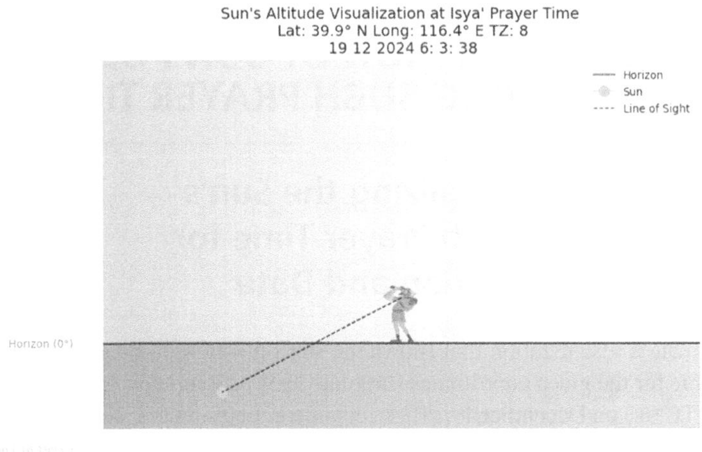

FIGURE 10.5 Visualization of sun position during Isya' prayer time.

```
# Add altitude scale
ax.axhline(0, color="black", linestyle="-",
linewidth=1)
ax.text(-0.5, 0, "Horizon (0°)", va="center",
ha="right", fontsize=10, color="green")
ax.text(-0.5, sun_y-1, f"Sun Position ({altitude_
angle:.4f}°)", va="center", ha="right", fontsize=10,
color="orange")

# Adjust the plot
ax.set_xlim(0, 10)
ax.set_ylim(-1.5, 5)
# Add labels and legend
ax.axis("off")
ax.legend(loc="upper right")
ax.set_title(f"Sun's Altitude Visualization at Isya'
Prayer Time\n Lat: {lat_location}° N Long: {long_
location}° E TZ: {timezone}\n {day} {month} {year}
{degrees}: {minutes}: {seconds}", fontsize=14)

plt.show()
```

The result of the code implementation above is shown in Figure 10.5.

VISUALIZATION OF SUN POSITION DURING SUBH PRAYER TIME

Exercise 6: Visualizing the Sun's Altitude at Subh Prayer Time for the Given Location and Date

Create a visualization that shows the sun's position in the sky at Subh prayer time for the given coordinates (latitude: 39.9°N, longitude: 116.4°E, time zone: UTC+8) on December 19, 2024, for solar depression degree of 15. First, determine the Subh prayer time

```
#Import Necessary Function
from skyfield.api import load
from skyfield.api import N, S, E, W, wgs84
from skyfield import almanac
import math

#Load Ephemeris Data and Planet Objects
ts = load.timescale()
eph = load('de440s.bsp')
planets = load('de440s.bsp')
earth = planets['earth']
sun = planets['sun']

# Variable Input

lat_location = 39.9
long_location = 116.4
timezone = 8
day = 19
month = 12
year = 2024
ele = 100

location = earth + wgs84.latlon(lat_location, long_
location, elevation_m=ele)
t0 = ts.utc(year, month, day)
t1 = ts.utc(year, month, day + 1)
```

```
from skyfield.units import Angle
from numpy import arccos
from skyfield.earthlib import refraction

altitude_m = ele
earth_radius_m = 6378136.6
side_over_hypotenuse = earth_radius_m / (earth_
radius_m + altitude_m)
h = Angle(radians=-arccos(side_over_hypotenuse))
solar_radius_degrees = 16 / 60
r = refraction(0.0, temperature_C=15.0,
pressure_mbar=1030.0)

t, y = almanac.find_risings(location, sun, t0, t1,
horizon_degrees=-r + h.degrees - solar_radius_degrees)
h, m, s = t.utc.hour, t.utc.minute, t.utc.second

subh_angle = 16
elevation_correction = 0.0293 * math.sqrt(ele)
subh_angle_actual = -subh_angle - elevation_correction

# Start with hour

test = 1
while True:

    # Calculate the Solar Altitude
    sun_astro = location.at(ts.utc(year, month, day, h,
    m, s)).observe(sun)
    sun_alt, _, _ = sun_astro.apparent().altaz() # Get
    the altitude of the sun

    if sun_alt.degrees >= 0:
      break
    if test > 24:
      break

    if sun_alt.degrees <= subh_angle_actual:
        break # Exit the loop if the solar altitude
        located below -18 degree
    h -= 1

# Once the condition is met for hours, move to minutes
h_subh = h + 1
test =1
```

```
while True:
   # Calculate the Solar Altitude
   sun_astro = location.at(ts.utc(year, month, day,
   h_subh, m, s)).observe(sun)
   sun_alt, _, _ = sun_astro.apparent().altaz() # Get
   the altitude of the sun
   if sun_alt.degrees >= 0:
     break
   if test > 1440:
     break

   if sun_alt.degrees <= subh_angle_actual:
       break # Exit the loop if the solar altitude
       located below -18 degree
   m -= 1

   # Increment time in minutes
m_subh = m + 1

# Once the condition is met for minutes, move to
seconds
test = 1
while True:
   # Calculate the Solar Altitude
   sun_astro = location.at(ts.utc(year, month, day,
   h_subh, m_subh, s)).observe(sun)
   sun_alt, _, _ = sun_astro.apparent().altaz() # Get
   the altitude of the sun
   if sun_alt.degrees >= 0:
     break
   if test > 86400:
     break

   if sun_alt.degrees <= subh_angle_actual:
       break # Exit the loop if the solar altitude
       located below -18 degree
   s -= 1

     # Increment time in seconds

   s_subh = s +1

subh_time = float(h_subh + (m_subh) / 60 + s_subh /
3600 + timezone)
```

```
subh_time %= 24 # Ensure 24-hour clock format
subh_time = float(subh_time)
degrees = int(subh_time)
decimal_part = subh_time - degrees
minutes_total = decimal_part * 60
minutes = int(minutes_total)
seconds = round((minutes_total - minutes) * 60)

if sun_alt.degrees >= 0 or test >86400:
    subh = "Subuh Does Not Occur"
else:
    subh = f"Subuh Occurs at {degrees}° {minutes}'
    {seconds}""

print(subh)
```

Determine the sun altitude of Subh

```
# Local time values
h = degrees
m = minutes
s = seconds

# Adjust local time to UTC
h_utc = h - timezone

# Compute observation time using Skyfield
sun_astro = location.at(ts.utc(year, month, day,
h_utc, m, s)).observe(sun)
sun_alt, _, _ = sun_astro.apparent().altaz()

print(sun_alt)
-16deg 11' 25.1"
```

Determine and visualize the position of the sun during Subh prayer time.

```
# 1. Import Library
import matplotlib.pyplot as plt
import numpy as np
from PIL import Image
import requests
import io # Import io to handle the image data in
memory
```

```
# 2. Set Observer and Sun Positions
observer_x, observer_y = 5, 0 # Observer's position on
the horizon
altitude_angle = sun_alt.degrees # Angle in degrees
(negative for below horizon)
distance_to_sun = 3 # Arbitrary horizontal distance to
the Sun

# 3. Calculate the Sun's position based on the angle
sun_x = observer_x - distance_to_sun # Place the Sun
to the left of the observer
sun_y = observer_y + np.tan(np.radians(altitude_
angle)) * distance_to_sun # Calculate vertical
position

# 4. Plot the Visualization
fig, ax = plt.subplots(figsize=(10, 6))

# Draw the horizon (straight line)
ax.plot([0, 10], [0, 0], color="green", linewidth=2,
label="Horizon")

# Add stick figure to the plot using direct download
link from Google Drive URL
image_url = "https://drive.google.com/uc?export
=download&id=1T7pLZNW6dF9PKxdOZ9UXZDbA84teRSdx"
# Download the image content from the URL
response = requests.get(image_url)
response.raise_for_status() # Raise an exception for
bad status codes
# Open the image from the downloaded content
stick_figure = Image.open(io.BytesIO(response.
content))

stick_x, stick_y = 5, 0 # Position of stick figure (on
the horizon)
ax.imshow(stick_figure, extent=(stick_x - 0.3, stick_x
+ 0.3, stick_y, stick_y + 1))

# Add the Sun as an orange dot
ax.plot(sun_x, sun_y, marker="o", color="orange",
markersize=10, label="Sun")

# Add sky gradient (blue)
sky_gradient = np.linspace(1, 0, 256).reshape(1, -1)
sky_gradient = np.vstack((sky_gradient, sky_gradient))
```

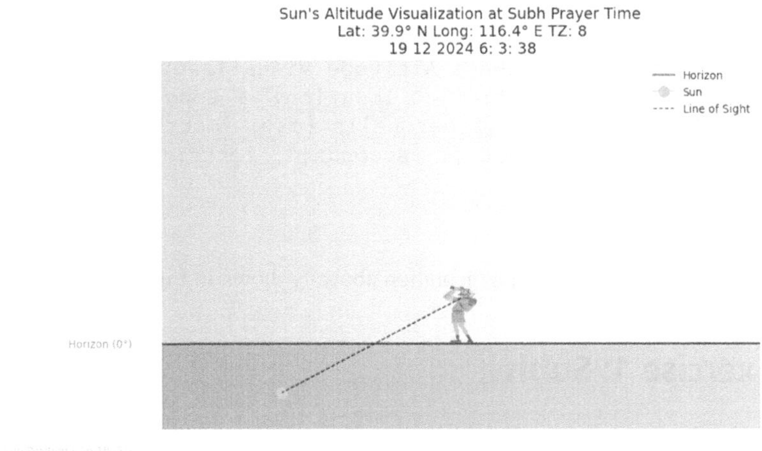

FIGURE 10.6 Visualization of sun position during Subh prayer time.

```
ax.imshow(sky_gradient, extent=[0, 10, -1.5, 5],
cmap=‘Blues’, alpha=0.3, aspect=‘auto’)

# Add ground (green)
ground = plt.Rectangle((0, -1.5), 10, 1.5,
color=‘darkgreen’, alpha=0.3)
ax.add_patch(ground)

# Draw the line of sight (dashed line)
ax.plot([observer_x, sun_x], [observer_y + 0.8,
sun_y], color=“black”, linestyle=“--”, linewidth=1,
label=“Line of Sight”)

# Add altitude scale
ax.axhline(0, color=“black”, linestyle=“-”,
linewidth=1)
ax.text(-0.5, 0, “Horizon (0°)”, va=“center”,
ha=“right”, fontsize=10, color=“green”)
ax.text(-0.5, sun_y-1, f”Sun Position ({altitude_
angle:.4f}°)”, va=“center”, ha=“right”, fontsize=10,
color=“orange”)

# Adjust the plot
ax.set_xlim(0, 10)
ax.set_ylim(-1.5, 5)
# Add labels and legend
```

```
ax.axis("off")
ax.legend(loc="upper right")
ax.set_title(f"Sun's Altitude Visualization at Subh
Prayer Time\n Lat: {lat_location}° N Long: {long_
location}° E TZ: {timezone}\n {day} {month} {year}
{degrees}: {minutes}: {seconds}", fontsize=14)

plt.show()
```

The result of the code implementation above is shown in Figure 10.6.

Exercise 1: Subh

Using the coordinates of Tokyo, which are 35.6894° North, 139.6916° East, elevation 100 m, and a time zone of GMT+9, calculate the Subuh (Fajr) prayer time for the date 27 December 2025 using a solar depression angle of −18 degrees, and visualize the sun's altitude when it reaches this angle before sunrise.

Exercise 2: Syuruk

Using the coordinates of Tokyo, which are 35.6894° North, 139.6916° East, elevation 100 m, and a time zone of GMT+9, calculate the Syuruk (sunrise) time for the date 27 December 2025 and visualize the sun's altitude curve around sunrise.

Exercise 3: Zuhur

Using the coordinates of Tokyo, which are 35.6894° North, 139.6916° East, elevation 100 m, and a time zone of GMT+9, calculate the Zuhur (prayer time) for the date 27 December 2025, and visualize the sun's altitude at its highest point on that day.

Exercise 4: Asar

Using the coordinates of Tokyo, which are 35.6894° North, 139.6916° East, elevation 100 m, and a time zone of GMT+9, calculate the Asar prayer time for the date 27 December 2025 using both the standard shadow ratio (1×) and the Hanafi method (2×), and visualize the sun's altitude now of each Asar time.

Exercise 5: Maghrib

Using the coordinates of Tokyo, which are 35.6894° North, 139.6916° East, elevation 100 m, and a time zone of GMT+9, calculate the Maghrib prayer time for the date 27 December 2025 based on the time of sunset and visualize the sun's altitude as it crosses the horizon.

Exercise 6: Isya'

Using the coordinates of Tokyo, which are 35.6894° North, 139.6916° East, elevation 100 m, and a time zone of GMT+9, calculate the Isya' (Isha) prayer time for the date 27 December 2025 using a solar depression angle of −18 degrees, and visualize the sun's altitude as it reaches that angle after sunset.

Lunar Crescent Observation Data Visualization 11

Matplotlib is a plotting library for the Python programming language and its numerical mathematics extension, NumPy. It provides an object-oriented API for embedding plots into applications using general-purpose GUI toolkits like Tkinter, wxPython, Qt, or GTK. There is also a procedural "pylab" interface based on a state machine (like OpenGL), designed to closely resemble MATLAB, though its use is discouraged. SciPy uses Matplotlib. Matplotlib was originally written by John D. Hunter. Since then, it has developed an active community and is distributed under a BSD-style license. Michael Droettboom was appointed as the lead developer of Matplotlib shortly before John Hunter's passing in August 2012 and was later joined by Thomas Caswell. Matplotlib can be used to create maps, plot various types of graphs, and is highly flexible for customization. Therefore, in this class, we will learn how to use Matplotlib for generating visualizations of crescent moon (*hilal*) data.

FIRST PRACTICE: PENANG MALAYSIA

A. Horizon Generation

The horizon generation must cover 360 degrees in azimuth and 180 degrees in altitude. This ensures that the generated horizon can visualize the visibility data of the crescent moon (*hilal*) across various locations and dates. The determination for horizon generation is as follows:

```
import matplotlib.pyplot as plt
import numpy as np

fig, ax = plt.subplots(figsize=(20,10))
```


DOI: 10.1201/9781003649120-11

```
horizon_angles = np.linspace(0, 360, 720)
horizon_altitudes = np.zeros_like(horizon_angles)
ax.plot(horizon_angles, horizon_altitudes,
color='black', linestyle='-', linewidth=1)
```

The result of the above programming is shown in Figure 11.1.

The generated line represents the observer's horizon line.

B. Solar Position Visualization and Graph Labelling

To display the sun's position (represented in yellow), the following programming steps are implemented:

```
# Sun Position Determination
sun_az = 287
sun_alt = -1

moon_az = 283
moon_alt = 12.00474
daz = abs(sun_az-moon_az)
arcl = 13.03075
Location = "Penang, Malaysia"
day = 12
month = 1
year = 2024
month_name = calendar.month_name[month]

The graph is annotated using these key methods:
ax.set_xlabel('Azimuth (degrees)')
ax.set_ylabel('Altitude (degrees)')
ax.set_title(' The Position of the Moon and Sun During
Observation ')
```

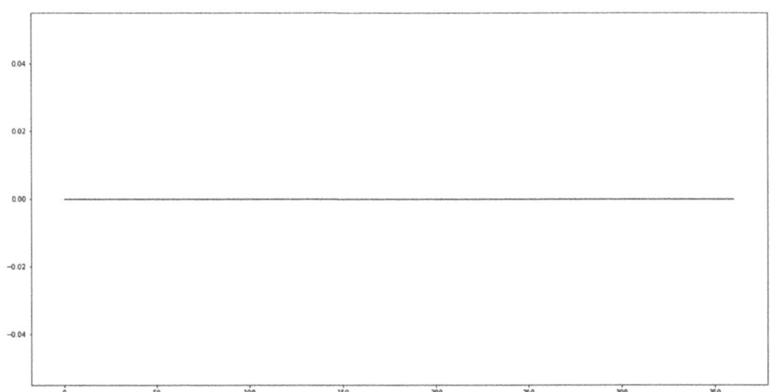

FIGURE 11.1 Observer horizon line.

With this, the altitude and azimuth have been labelled for easier interpretation. Typically, we observe the sun and the central reference line aligned at the midpoint. To ensure the black horizon line and the sun are centered, the following adjustments are made (Figure 11.2):

```
xlim_max = max(sun_az - (daz * 2), sun_az + (daz * 2))
xlim_min = min(sun_az - (daz * 2), sun_az + (daz * 2))

ax.set_xlim((xlim_min, xlim_max))
ax.set_ylim((sun_alt - 2), (moon_alt + 5))
```

C. Next is to display the moon's position. The moon's shape uses a crescent moon image. Please download the crescent moon position file from Google image search with a transparent background. Upload the crescent moon image to Google Drive with public access. The programming to determine the crescent moon's position is as follows:

```
from PIL import Image
import requests
import io # Import io to handle the image data in
memory
from matplotlib.offsetbox import OffsetImage,
AnnotationBbox
# Add the crescent moon image as a marker
opposite = moon_alt - sun_alt
adjacent = (moon_az - sun_az)
```

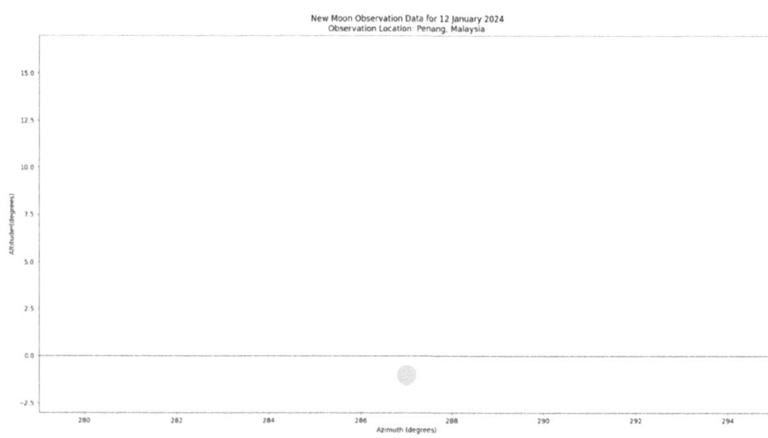

FIGURE 11.2 Observer horizon line with sun.

```
# Calculate the angle using the tangent function (TOA)
angle_rad = math.atan2(opposite, adjacent)

# Convert the angle to degrees
angle_degrees = (math.degrees(angle_rad))

# Load the crescent moon image
# Modified the Google Drive URL to get the direct
download link
image_url = "https://drive.google.com/uc?export
=download&id=1ZFbJ5pWYv3ZE4SY50Rme7w8iejzKRik4"

# Download the image content from the URL
response = requests.get(image_url)
response.raise_for_status() # Raise an exception for
bad status codes
# Open the image from the downloaded content
crescent_img = Image.open(io.BytesIO(response.
content))

# Rotate the crescent moon image based on the
calculated angle
rotated_img = crescent_img.rotate(angle_degrees)

# Add the crescent moon image as a marker
imagebox = OffsetImage(rotated_img, zoom=0.03) #
Adjust zoom as needed
# Using the correct moon position variables moon_az
and moon_alt
ab = AnnotationBbox(imagebox, (moon_az, moon_alt),
frameon=False)
ax.add_artist(ab)

plt.show() # Added this line to display the plot
```

Thus, the crescent moon's position relative to the sun can be clearly illustrated (Figure 11.3).

D. Sky Background Generation

The next step involves creating the celestial background. The implementation requires these steps:

```
from matplotlib.colors import LinearSegmentedColormap
# Import LinearSegmentedColormap
```

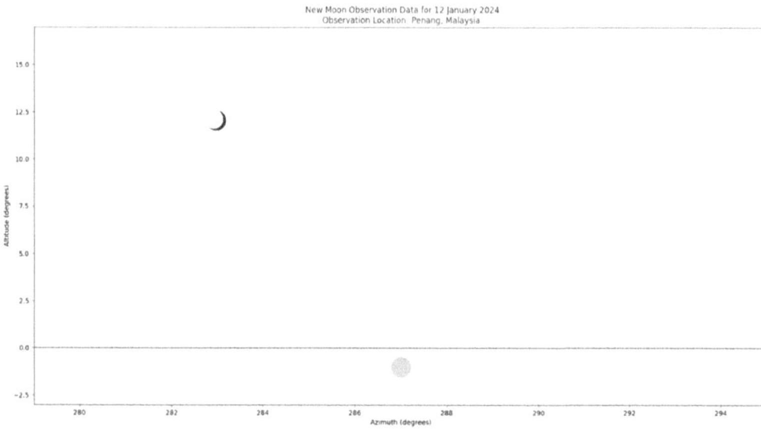

FIGURE 11.3 Observer horizon line with sun and moon.

```
# Sky Background Generation
sky = LinearSegmentedColormap.from_list('sky',
['blue','white', 'yellow', 'orange'])
extent = ax.get_xlim() + ax.get_ylim() # Adjusted to
use current xlim and ylim
ax.imshow([[0, 0], [1, 1]], cmap=sky,
interpolation='bicubic', extent=extent)
```

E. Next is to display the MABIMS criteria on the visualization. First, we'll visualize the elongation criterion. On this plot, elongation is represented as a radial distance from the sun's center. The implementation code is as follows (Figure 11.4):

```
from matplotlib.patches import Arc # Import the Arc
class
# Visualization of Elongation Criteria
sun_az_degrees = sun_az
sun_alt_degrees = sun_alt
moon_az_degrees = moon_az
moon_alt_degrees = moon_alt

kriteria_elongasi = 6.4
circle_radius = kriteria_elongasi

# Sun center coordinates
sun_center_x = sun_az
sun_center_y = sun_alt
```

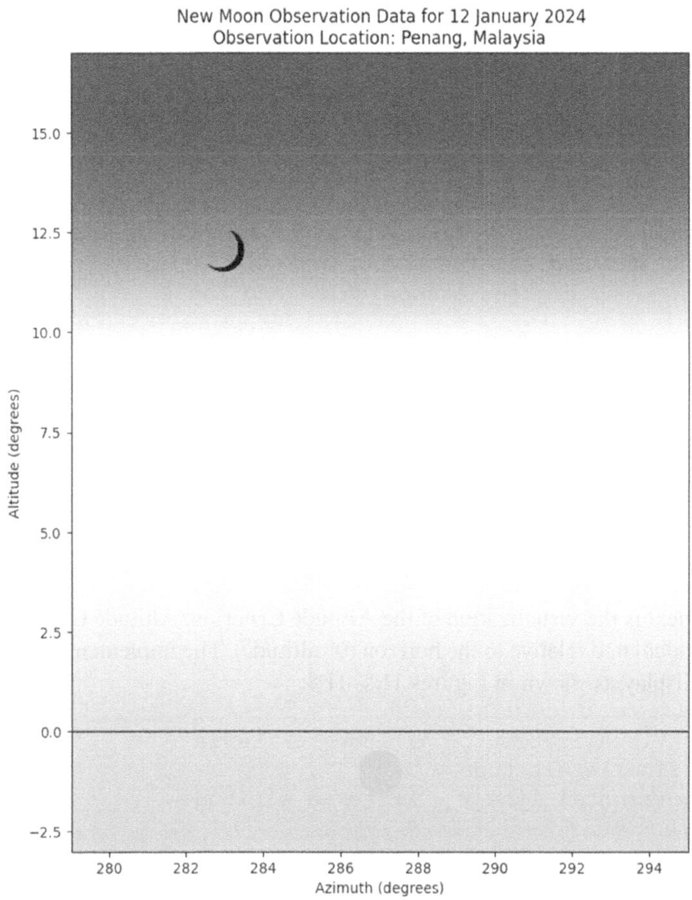

FIGURE 11.4 Visualization of observer horizon with sun and moon position.

```
# Semicircle parameters
radius = kriteria_elongasi
x1 = sun_az-kriteria_elongasi
x2 = sun_az+kriteria_elongasi

# Calculate the starting and ending angles of the
semicircle

arccos_value_start_angle = np.arccos(((x1 - sun_
center_x) / radius))
arccos_value_end_angle = np.arccos(((x2 - sun_
center_x) / radius))
```

```
if np.isnan(np.arccos((arccos_value_start_angle))):
    start_angle = 0.000001
else:
    start_angle = 180
    - np.degrees(arccos_value_start_angle)

if np.isnan(np.arccos((arccos_value_end_angle))):
    end_angle = 180+0.000001
else:
    end_angle = 180
    - np.degrees(arccos_value_end_angle)

print(start_angle,end_angle)
# Create the semicircle patch
semicircle_patch = Arc((sun_center_x, sun_center_y), 2
* radius, 2 * radius, theta1=start_angle,
theta2=end_angle,
                fill=False, color='blue', linestyle='--')
# Add the semicircle patch to the plot
ax.add_patch(semicircle_patch)
```

F. Next is the visualization of the Altitude Criterion. Altitude Criteria are calculated relative to the horizon (0° altitude). The implementation will display as shown in Figures 11.5–11.8:

```
# Visualization of Altitude Criteria
kriteria_altitude = 3
horizontal_line_y = kriteria_altitude
x1 = 0
x2 = sun_az-kriteria_elongasi
ax.hlines(y=horizontal_line_y,xmin=x1,
xmax=x2,color='red', linestyle='--')

x11 = sun_az+kriteria_elongasi
x22 = 360
ax.hlines(y=horizontal_line_y,xmin=x11,
xmax=x22,color='red', linestyle='--')
```

G. Add Logo

```
# Load the Logo
# Modified the Google Drive URL to get the direct
download link
logo_url = "https://drive.google.com/uc?export
=download&id=1_HEq0C3bv33i54HlbrJKD4Zmdxap8Q8a"
```

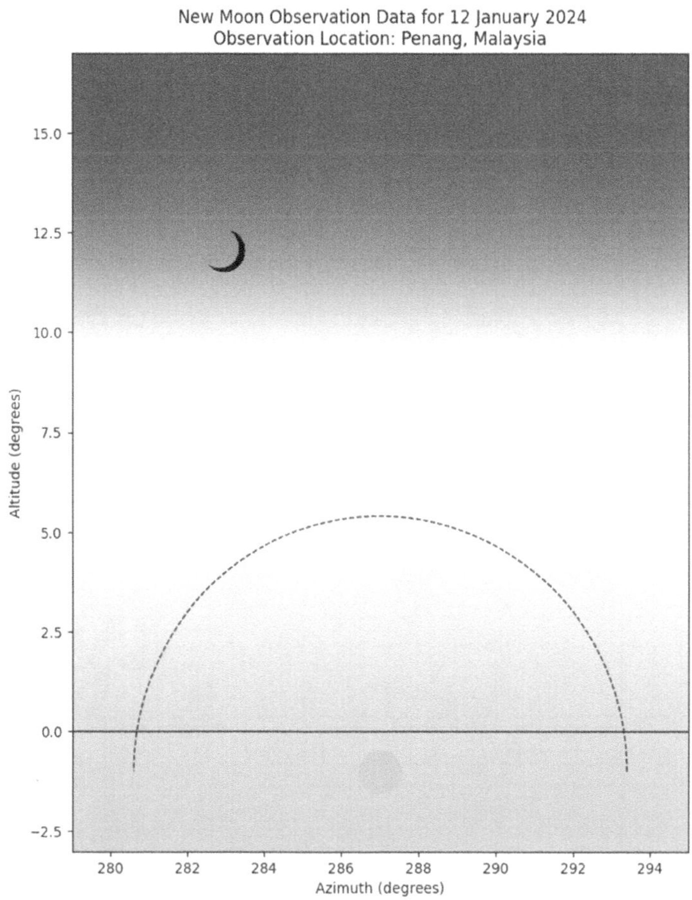

FIGURE 11.5 Visualization of observer horizon with sun and moon position with elongation criterion.

```
# Download the image content from the URL
response = requests.get(logo_url)
response.raise_for_status() # Raise an exception for
bad status codes
# Open the image from the downloaded content
logo_img = Image.open(io.BytesIO(response.content))

# Add the logo to the plot (bottom right corner)
logo_box = OffsetImage(logo_img, zoom=0.1) # Adjust
zoom as needed
```

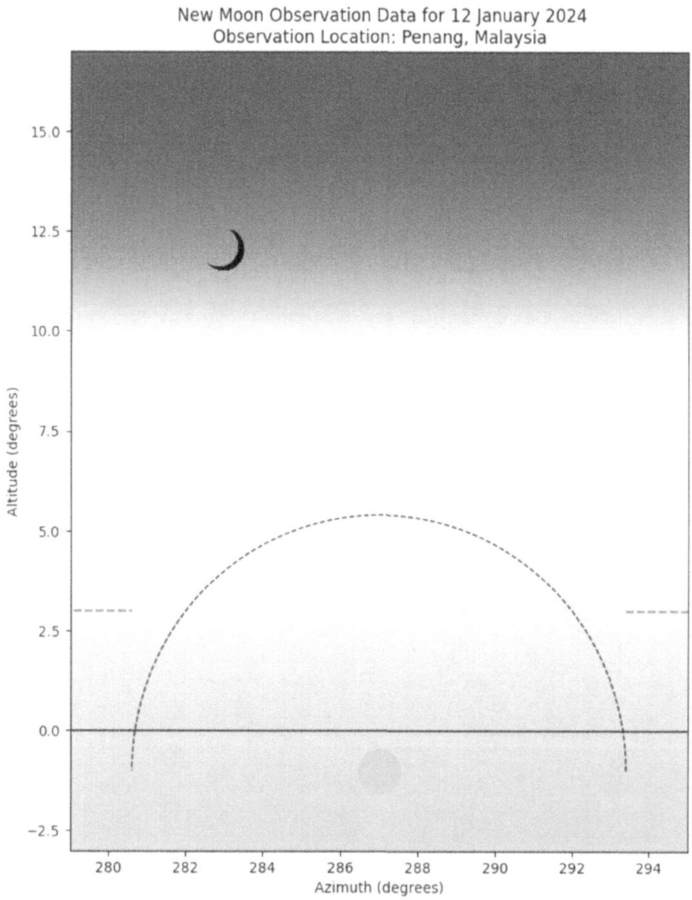

FIGURE 11.6 Visualization of observer horizon with sun and moon position with elongation and altitude criterion.

```
logo_ab = AnnotationBbox(logo_box, (xlim_max - 2.5,
sun_alt - 1), frameon=False) # Position adjusted based
on plot limits
ax.add_artist(logo_ab)
```

H. Plot the Moon's Altitude Line and Elongation

```
# Altitude Line
ax.vlines(x=moon_az, ymin=0, ymax=moon_alt,
color='blue', linestyle='--')
```

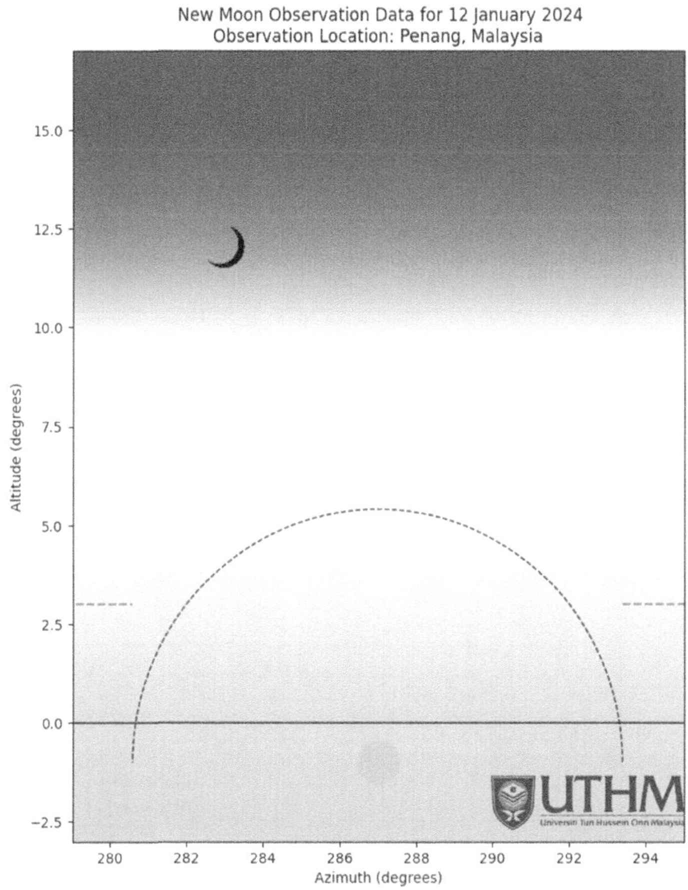

FIGURE 11.7 Visualization of observer horizon with sun and moon position with elongation and altitude criterion, added with logo.

```
ax.text(moon_az+0.3, moon_alt/2, f'Moon Altitude :
{moon_alt:.2f}', color='blue', fontsize=10, ha='left')

# Elongation Line
ax.plot([moon_az, sun_az], [moon_alt, sun_alt],
color='green', linestyle='--')
ax.text(sun_az-2, arcl/2-1, f'Elongation: {arcl:.2f}',
color ='green', fontsize=10, ha='left')

plt.show() # Added this line to display the plot
```

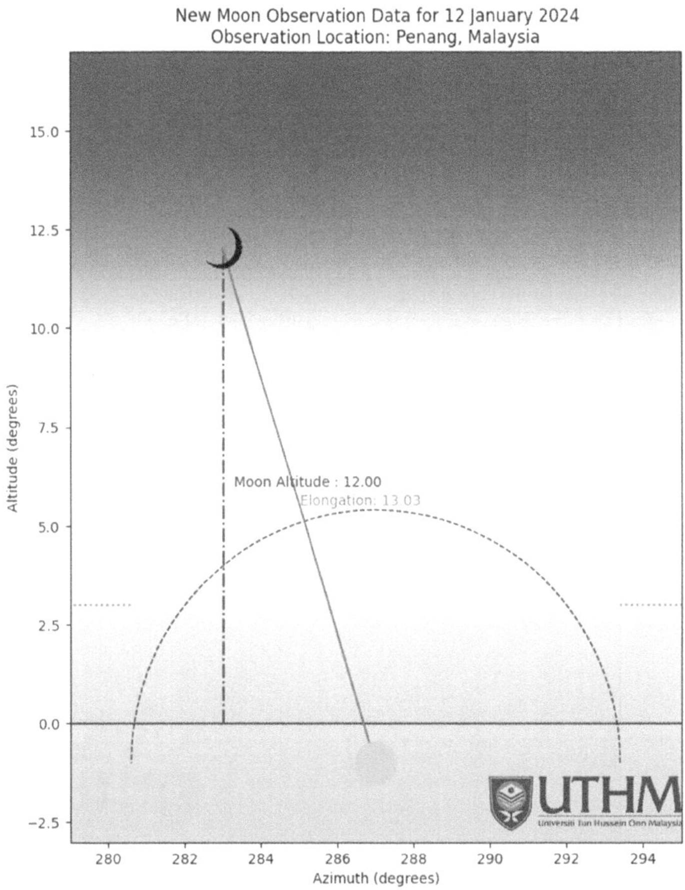

FIGURE 11.8 Visualization of observer horizon with sun and moon position with elongation and altitude criterion, added with logo and dash line.

SECOND PRACTICE: BANDA ACEH INDONESIA

Exercise 1

Visualize new moon observation data for May 27, 2025, on Observation Location: Banda Aceh, Indonesia, Lat: 5.548290 N, 95.323753 East, UTC+7.

```
# Install required libraries
!pip install skyfield # Astronomical calculations
!pip install numpy # Numerical operations
!pip install scipy # Scientific computing
!pip install matplotlib # Plotting library
!pip install tabulate # Pretty-print tables

# Import necessary modules
from skyfield import almanac
from skyfield.api import Topos,load
from skyfield import api
import numpy as np
from skyfield.api import N, S, E, W, load, wgs84
from skyfield.api import Topos, load,
Angle,GREGORIAN_START
import math
from scipy.ndimage import rotate
import calendar
from tabulate import tabulate
from matplotlib.patches import Arc
import matplotlib.pyplot as plt
from matplotlib.colors import LinearSegmentedColormap
import matplotlib.image as mpimg
from matplotlib.offsetbox import OffsetImage,
AnnotationBbox
from PIL import Image
import requests
import io # Import io to handle the image data in
memory

# Load planetary ephemeris data
planets = load('de421.bsp')
earth = planets['earth']
sun = planets['sun']
moon = planets['moon']
h_maghrib = 0 # Maghrib hour
m_maghrib = 0 # Maghrib minute

# Initialize time scale and ephemeris
ts = load.timescale()
eph = api.load('de421.bsp')

# Set observation location and time parameters
Lokasi = "Banda Aceh, Indonesia"
lat_titik1 = 5.548290 # Latitude of observation point
```

```
lon_titik1 = 95.323753 # Longitude of observation
point
tz = 7 # Timezone offset (UTC+7)
year = 2025 # Year of observation
month = 5 # Month of observation
day = 27 # Day of observation

# Create observer location object
location_titik1 = wgs84.latlon(lat_titik1 * N, lon_
titik1 * E)
observer_titik1 = eph['Earth'] + location_titik1
print(observer_titik1)

def settime(year, month, day, observer_titik, x):
  """Calculate setting time of celestial body with
  timezone adjustment"""
  # Set time range for search (current day to next day)
  t0 = ts.utc(year,month,day)
  t1 = ts.utc(year, month,day+1)

  # Find setting time of celestial body x
  t,y = almanac.find_settings(observer_titik, x, t0, t1)

  # Extract UTC time components
  h_set_transit_notz,m_set_transit_notz,s_set_transi
  t_notz = (int(t.utc.hour)),(int(t.utc.minute)),(i
  nt(t.utc.second))

  # Convert to decimal hours and add timezone offset
  time_set = (h_set_transit_notz + ((m_set_transit_
  notz) / 60 + s_set_transit_notz / 3600))+tz

  # Convert back to hours, minutes, seconds
  h_set, d = divmod(time_set, 1)
  h_set = int(h_set)
  m_set, s = divmod(d * 60, 1)
  m_set = int(m_set+1+4/60) # Add small adjustment
  s_set = int(s * 60)
  # Handle overflow in seconds
  if s_set >= 60:
      m_set += s_set // 60
      s_set = s_set % 60

  # Handle overflow in minutes
  if m_set >= 60:
```

```
        h_set += m_set // 60
        m_set = m_set % 60

    # Ensure hours wrap around a 24-hour clock
    h_set = h_set % 24
    waktu_terbenam = f"{h_set}:{m_set:02}:{s_set:02}"

    return waktu_terbenam

# Calculate sunset time
Objek = 'Matahari'
sun_set = settime(year, month, day, observer_titik1,
sun)

# Calculate moonset time
Objek = 'Bulan'
moon_set = settime(year, month, day, observer_titik1,
moon)

# Set Julian calendar cutoff for historical dates
ts.julian_calendar_cutoff = GREGORIAN_START
location = Topos(latitude_degrees=lat_titik1,
longitude_degrees=lon_titik1, elevation_m=1)

def settime(year, month, day, observer_titik, x):
    """Alternative version that returns hours and
    minutes separately"""
    # Similar to previous function but returns numeric
    values
    t0 = ts.utc(year,month,day)
    t1 = ts.utc(year, month,day+1)
    t,y = almanac.find_settings(observer_titik, x, t0,
    t1)
    h_set_transit_notz,m_set_transit_notz,s_set_transi
    t_notz = (int(t.utc.hour)),(int(t.utc.minute)),(i
    nt(t.utc.second))
    time_set = (h_set_transit_notz + ((m_set_transit_
    notz) / 60 + s_set_transit_notz / 3600))

    h_set, d = divmod(time_set, 1)
    h_set = int(h_set)
    m_set, s = divmod(d * 60, 1)
    m_set = int(m_set+1+4/60)
    s_set = int(s * 60)
```

```
    if s_set >= 60:
        m_set += s_set // 60
        s_set = s_set % 60

    if m_set >= 60:
        h_set += m_set // 60
        m_set = m_set % 60

    h_set = h_set % 24
    waktu_terbenam = f"{h_set}:{m_set:02}:{s_set:02}"

    return h_set,m_set

# Get sunset time in numeric format
hsunset,msunset = settime(year, month, day, observer_
titik1, sun)

# Create observation location object
boston = earth + Topos(latitude_degrees=lat_titik1,
longitude_degrees=lon_titik1, elevation_m=0)

# Calculate sun position at sunset
sun_astro = boston.at(ts.utc((year), (month), (day),
(hsunset), (msunset))).observe(sun)
sun_app = sun_astro.apparent()
sun_alt, sun_az, sun_distance = sun_app.altaz()

# Calculate moon position at sunset
moon_astro = boston.at(ts.utc((year), (month), (day),
(hsunset), (msunset))).observe(moon)
moon_app = moon_astro.apparent()
moon_alt, moon_az, moon_distance = moon_app.altaz()

# Calculate differences between moon and sun positions
beza_altitud_bulan_matahari = abs(moon_alt.degrees-
sun_alt.degrees)
daz = abs(moon_az.degrees-sun_az.degrees)
str_date = f'{day}/{month}/{year}'

# Store observation data
altitud_bulan = moon_alt.degrees
elongasi = sun_app.separation_from(moon_app).degrees

# Prepare data for table display
data = [["Date", str_date],
```

```
    ["Location", Lokasi],
    ["Sunset Time", sun_set],
    ["Moonset Time", moon_set],
    ["Moon Altitude", altitud_bulan],
    ["Elongation", elongasi],
]

col_names = ["Moon-Sun Data", "Value"]

# Create formatted table
data_table = (tabulate(data, headers=col_names))

# Create visualization figure
fig, ax = plt.subplots(figsize=(20,10))

# Draw horizon line
horizon_angles = np.linspace(0, 360, 720)
horizon_altitudes = np.zeros_like(horizon_angles)
ax.plot(horizon_angles, horizon_altitudes,
color='black', linestyle='-', linewidth=1)

# Get sun and moon positions in degrees
sun_az = sun_az.degrees
sun_alt = sun_alt.degrees
moon_az = moon_az.degrees
moon_alt = moon_alt.degrees

arcl = elongasi
Location = Lokasi
month_name = calendar.month_name[month]

# Plot sun position
ax.scatter(sun_az, sun_alt, color='orange',
label='Sun', zorder=10, s=900)
ax.set_xlabel('Azimuth (degrees)')
ax.set_ylabel('Altitude (degrees)')
ax.set_title(f'New Moon Observation Data for {day}
{month_name} {year}\nObservation Location:
{Location}')

# Set plot limits based on sun-moon positions
xlim_max = max(sun_az - (daz * 2), sun_az + (daz * 2))
xlim_min = min(sun_az - (daz * 2), sun_az + (daz * 2))
ax.set_xlim((xlim_min, xlim_max))
ax.set_ylim((sun_alt - 2), (moon_alt + 5))
```

```
# Calculate angle for moon crescent orientation
opposite = moon_alt - sun_alt
adjacent = (moon_az - sun_az)
angle_rad = math.atan2(opposite, adjacent)
angle_degrees = (math.degrees(angle_rad))

# Load and rotate crescent moon image
image_url = "https://drive.google.com/uc?export
=download&id=1ZFbJ5pWYv3ZE4SY50Rme7w8iejzKRik4"
response = requests.get(image_url)
response.raise_for_status()
crescent_img = Image.open(io.BytesIO(response.
content))
rotated_img = crescent_img.rotate(angle_degrees)

# Add rotated moon image to plot
imagebox = OffsetImage(rotated_img, zoom=0.03)
ab = AnnotationBbox(imagebox, (moon_az, moon_alt),
frameon=False)
ax.add_artist(ab)

# Create sky background gradient
sky = LinearSegmentedColormap.from_list('sky',
['blue','white', 'yellow', 'orange'])
extent = ax.get_xlim() + ax.get_ylim()
ax.imshow([[0, 0], [1, 1]], cmap=sky,
interpolation='bicubic', extent=extent)

# Visualization of elongation criteria (6.4°
semicircle)
kriteria_elongasi = 6.4
circle_radius = kriteria_elongasi
sun_center_x = sun_az
sun_center_y = sun_alt
radius = kriteria_elongasi
x1 = sun_az-kriteria_elongasi
x2 = sun_az+kriteria_elongasi

# Calculate angles for semicircle
arccos_value_start_angle = np.arccos(((x1 - sun_
center_x) / radius))
arccos_value_end_angle = np.arccos(((x2 - sun_
center_x) / radius))

if np.isnan(np.arccos((arccos_value_start_angle))):
    start_angle = 0.000001
```

```
else:
    start_angle = 180
    - np.degrees(arccos_value_start_angle)

if np.isnan(np.arccos((arccos_value_end_angle))):
    end_angle = 180+0.000001
else:
    end_angle = 180
    - np.degrees(arccos_value_end_angle)

# Draw semicircle around sun position
semicircle_patch = Arc((sun_center_x, sun_center_y), 2
* radius, 2 * radius,
               theta1=start_angle, theta2=end_angle,
               fill=False, color='blue',
               linestyle='--')
ax.add_patch(semicircle_patch)

# Visualization of altitude criteria (3° line)
kriteria_altitude = 3
horizontal_line_y = kriteria_altitude
x1 = 0
x2 = sun_az-kriteria_elongasi
ax.hlines(y=horizontal_line_y, xmin=x1, xmax=x2,
color='red', linestyle='--')

x11 = sun_az+kriteria_elongasi
x22 = 360
ax.hlines(y=horizontal_line_y, xmin=x11, xmax=x22,
color='red', linestyle='--')

# Load and add logo
logo_url = "https://drive.google.com/uc?export
=download&id=1_HEq0C3bv33i54HlbrJKD4Zmdxap8Q8a"
response = requests.get(logo_url)
response.raise_for_status()
logo_img = Image.open(io.BytesIO(response.content))
logo_box = OffsetImage(logo_img, zoom=0.1)
logo_ab = AnnotationBbox(logo_box, (xlim_max - 2.5,
sun_alt - 1), frameon=False)
ax.add_artist(logo_ab)

# Draw moon altitude line and label
ax.vlines(x=moon_az, ymin=0, ymax=moon_alt,
color='blue', linestyle='--')
```

```
ax.text(moon_az+0.3, moon_alt/2, f'Moon Altitude:
{moon_alt:.2f}°', color='blue', fontsize=10,
ha='left')

# Draw elongation line and label
ax.plot([moon_az, sun_az], [moon_alt, sun_alt],
color='green', linestyle='--')
ax.text(sun_az+2, arcl/2-1, f'Moon Elongation:
{arcl:.2f}°', color='green', fontsize=10, ha='left')

# Determine visibility based on MABIMS criteria
if moon_alt >= kriteria_altitude and arcl >=
kriteria_elongasi:
    kenampakan = "Visible"
    nama_kriteria = "MABIMS Criteria Met"
else:
    kenampakan = "Not Visible"
    nama_kriteria = "MABIMS Criteria Not Met"

# Add information box
info_text = (f'Moon Data:\n'
          f'Altitude: {moon_alt:.2f}°\n'
          f'Elongation: {arcl:.2f}°\n'
          f'Visibility: {kenampakan}\n'
          f'Criteria: {nama_kriteria}')
ax.text(xlim_max-7, moon_alt+3.5, info_text,
     fontsize=12, ha='left', va='top',
     bbox=dict(facecolor='white', alpha=0.8))

# Add horizon and criteria labels
ax.text(xlim_min+1, 0.1, "Horizon", fontsize=10,
ha='left')
ax.text(xlim_min+1, kriteria_altitude+0.1, "MABIMS
2021 Criteria", fontsize=10, ha='left')

plt.show()

# Print data table
print('\n')
print(data_table)
```

The result of the above code is shown in Figure 11.9.

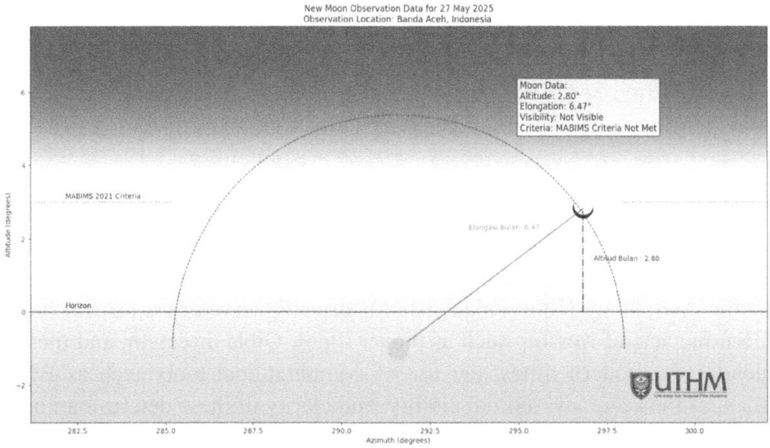

FIGURE 11.9 Visualization of sun position at Banda Acheh Indonesia.

Exercise 2

Visualize new moon observation data for June 25, 2025, on Observation Location: Singapore, Lat: 1.290270 N, 103.851959 E, UTC+8.

Exercise 3

Visualize new moon observation data for September 22, 2025, on Observation Location: Kuala Lumpur, Malaysia, Lat: 3.140853 N, 101.693207 E, UTC+8.

Exercise 4

Visualize new moon observation data for October 21, 2025, on Observation Location: Bandar Seri Begawan, Brunei, Lat: 4.890278 N, 114.942222 E, UTC+8.

Exercise 5

Visualize new moon observation data for December 20, 2025, on Observation Location: Naypyidaw, Myanmar, Lat: 19.7475 N, 96.115 E, UTC+6.5.

Conclusion

The fusion of Python programming and Islamic Astronomy presented in this textbook is not just a convergence of science and faith; it reflects the long-standing Islamic tradition that values precision, observation, and knowledge. Islamic Astronomy (Ilmu Falak) has historically served as a critical field to determine acts of worship such as prayer times, Qibla direction, and the Hijri calendar. In modern times, the use of computational tools such as Python enhances the accuracy, reproducibility, and clarity of these determinations.

Throughout this book, we have systematically explored how Python can be utilized to model and compute key astronomical phenomena relevant to Islamic practices. Beginning with fundamental programming skills, we built up to advanced computations including solar transit, shadow length analysis, and lunar crescent visibility prediction. We introduced the Skyfield library as a core tool, highlighting its ability to interface directly with astronomical data and simulate celestial motion with high precision.

In Chapters 5 through 7, we demonstrated how to calculate the direction of the Qibla and determine prayer times such as Zuhur, Asar, Maghrib, Isya', Syuruk, and Subh. These calculations were grounded in real astronomical parameters such as solar altitude, Equation of Time, and observer location. We emphasized the importance of considering elevation, atmospheric refraction, and solar declination, factors often overlooked in manual estimations. We also provided visualizations to help learners see the underlying logic and movement of the sun in relation to the Earth's surface.

In Chapter 8, we shifted to moonsighting, a subject of both scientific and sociological significance in the Muslim world. Here, readers were introduced to methods of determining the moon's altitude, age, elongation, and azimuth, all of which play crucial roles in validating the visibility of the lunar crescent (hilal). Through case studies and real-world examples, readers learned how to extract meaningful data that contributes to hilal visibility reports and the formulation of Hijri calendars.

Chapters 9 and 10 focused on visualization techniques that support understanding and teaching of Falak. With Python's graphical capabilities, learners could plot Qibla directions, solar azimuths, and sun positions at various prayer times. These visual outputs not only serve pedagogical purposes but also help in validating and communicating data effectively, particularly in official or community-level Falak determinations.

Finally, in Chapter 11, we explored lunar crescent observation data visualization using real observation scenarios from Malaysia and Indonesia. These practical examples tied together the themes of data, visibility, and Islamic calendar accuracy, showing that observation and computation must go hand-in-hand.

By combining modern computational skills with classical Islamic astronomical knowledge, this textbook aspires to produce a new generation of Muslim scholars, students, and professionals who are confident in applying scientific methods to religious obligations. This interdisciplinary skill set not only enhances one's understanding of Islamic rituals but also contributes to broader scientific literacy and critical thinking.

It is our hope that this book serves as a gateway for more Muslims, particularly students in higher education and religious institutions, to explore Islamic Astronomy as both an academic and spiritual pursuit. As we embrace the tools of modern science, we continue the legacy of great Muslim astronomers such as Al-Biruni, Al-Khwarizmi, and Al-Tusi, who exemplified the harmony between reason, revelation, and observation.

May this book inspire a deeper love for knowledge, a renewed commitment to precision in worship, and a stronger integration of technology in the service of our faith and community.

Wallāhu a'lam

References

Abas, A.-M., Safiai, M. H., Hasan, S. A., Azam, A. I., & Hussaini, R. (2022). Penentuan Waktu Ibadah Solat dan Puasa di Bangunan Pencakar Langit [Determining the times for prayer and fasting on skyscraper building]. *BITARA International Journal of Civilizational Studies and Human Sciences*, *6*(1), 53–59.

Adegoke, K. A. (2013). Ikhtilaf on moon sighting among Nigerian Muslims within the framework of Shari'ah. *Journal of Arabic and Islamic Studies*, *7*(2), 140–156.

Adegoke, K. A. (2017). Ādam al-Ilūrī and neo-ijtihād: An examination of his legal views on Ramadān fasting in Nigeria. *Journal of Arabic And Islamic Studies*, *20*(1), 11–25.

Ali, M. (2015). Is the British weather anti-Islamic? Prayer times, the ulama and application of the shari'a. *Contemporary Islam*, *9*(2), 171–187. https://doi.org/10.1007/s11562-014-0318-7

Al-Rajab, M., Loucif, S., & Al Risheh, Y. (2023). Predicting new crescent moon visibility applying machine learning algorithms. *Scientific Reports*, *13*(1), 6674. https://doi.org/10.1038/s41598-023-32807-x

Amin, M. F. (2018). Global Rasdhul Qibla: The probability of four times in a year study. *Jurnal Penelitian*, 15(2) 175–188.

Asrin, A., Hapsari, G. I., & Mutiara, G. A. (2018). Development of Qibla direction cane for blind using interactive voice command. *2018 6th International Conference on Information and Communication Technology (ICoICT)*, 216–221. https://doi.org/10.1109/ICoICT.2018.8528769

Blank, J., & Deb, K. (2020). Pymoo: Multi-objective optimization in Python. *IEEE Access*, *8(1)*, 89497–89509. https://doi.org/10.1109/ACCESS.2020.2990567

Elmhamdi, A., Roman, M. T., Peñaloza-Murillo, M. A., Pasachoff, J. M., Liu, Y., Al-Mostafa, Z. A., Maghrabi, A. H., Oloketuyi, J., & Al-Trabulsy, H. A. (2024). Impact of the eclipsed sun on terrestrial atmospheric parameters in desert locations: A comprehensive overview and two events case study in Saudi Arabia. *Atmosphere*, *15*(1), 62. https://doi.org/10.3390/atmos15010062

Faid, M. S., Husien, N., Shariff, N. N. M., Ali, M. O., Hamidi, Z. S., Zainol, N. H., & Sabri, S. N. U. (2016). Monitoring the level of light pollution and its impact on astronomical bodies naked-eye visibility range in selected areas in Malaysia using the sky quality meter. *2016 International Conference on Industrial Engineering, Management Science and Application (ICIMSA)*, 1–6. https://doi.org/10.1109/ICIMSA.2016.7504020

Faid, M. S., Md Shariff, N. N., & Hamidi, Z. S. (2019). The risk of light pollution on sustainability. *ASM Science Journal*, *12*(Special Issue 2), 134–142.

Faid, M. S., Mohd Nawawi, M. S. A., & Mohd Saadon, M. H. (2023). *HilalPy: Analysis tool for lunar crescent visibility criterion*. ascl:2307.031-ascl:2307.031.

Faid, M. S., Mohd Nawawi, M. S. A., & Mohd Saadon, M. H. (2024). *Design, development and analysis of lunar crescent visibility criterion with Python*. CRC Press.

Faid, M. S., Mohd Nawawi, M. S. A., & Norman, M. P. (2021). Issue of sustainability on light pollution from the perspective of Maqasid Shariah. *Journal of Fatwa Management and Research*, *26*(2), 1–9. https://doi.org/10.33102/jfatwa.vol26no2.390

Faid, M. S., Nawawi, M. M., & Saadon, M. M. (2023). Analysis tool for lunar crescent visibility criterion based on integrated lunar crescent database. *Astronomy and Computing*, *45(1)*, 100752.

Faid, M. S., Nawawi, M. S. A. M., Wahab, R. A., & Ahmad, N. (2023). HilalPy: Software to analyse lunar sighting criteria. *Software Impacts*, *18*, 100593.

Faid, M. S., Mohd Nawawi, M. S. A., Mohd Saadon, M. H., Wahid, K., & Norman, P. (2025). Methods in determining new Hijri month: A thematic review from Islamic jurisprudence perspective. *Malaysian Journal of Syariah and Law*, *13*(1), 75–99. https://doi.org/10.33102/mjsl.vol13no1.687

Faid, M. S., Nahwandi, M. S., Nawawi, M. S. A. B. M., Zaki, N. B. A., & Saadon, M. H. M. (2022). Development of Qibla direction determinant using sun shadow. *Online Journal of Research in Islamic Studies*, *9*(1), 89–102.

Faid, M. S., Nawawi, M. S. A. M., Saadon, M. H. M., Ahmad, N., & Ali @ Mat Zin, A. (2022). Islamic historical review on the middle age lunar crescent visibility criterion. *Journal of Al-Tamaddun*, *17*(1). https://doi.org/10.22452/JAT.vol17no1.9

Faid, M. S., Shariff, N. N. M., Hamidi, Z. S., Kadir, N., Ahmad, N., & Wahab, R. A. (2018). Semi empirical modelling of light polluted twilight sky brightness. *Jurnal Fizik Malaysia*, *39*(2), 30059–30067.

Faid, M. S., Shariff, N. N. M., Hamidi, Z. S., Sabri, S. N. U., Zainol, N. H., Husien, N. H., & Ali, M. O. (2016). Monitoring the level of light pollution and its impact on astronomical bodies naked-eye visibility range in selected areas in Malaysia using sky quality meter. *Journal of Industrial Engineering and Management Science*, *2016*(1), 1–18. https://doi.org/10.13052/jiems2446-1822.2016.007

Faid, M. S., Shariff, N. N. M., Hamidi, Z. S., Wahab, R. A., Ahmad, N., Mohd Nawawi, M. S. A., & Nahwandi, M. S. (2024). Alteration of twilight sky brightness profile by light pollution. *Scientific Reports*, *14*(1), 26682.

Faid, M. S., Nawawi, M. S. A. M., Saadon, M. H. M., Wahab, R. A., Ahmad, N., Nahwandi, M. S., Ahmed, I., & Mohamed, I. (2024). Assessment and review of modern lunar crescent visibility criterion. *Icarus*, *412*, 115970.

Faid, M. S., Nawawi, M. S. A. M., Saadon, M. H. M., Nahwandi, M. S., Shariff, N. N. M., Hamidi, Z. S., Wahab, R. A., Norman, M. P., & Ahmad, N. (2023). Confirmation methodology for a lunar crescent sighting report. *New Astronomy*, *103*, 102063–102063. https://doi.org/10.1016/j.newast.2023.102063

Gharaybeh, M. (2025). Jurisprudential reliance on astronomical calculations in determining the beginnings of the Hijri month. In H. M. K. Al Naimiy, H. M. Elmehdi, & I. A. Shehadi (Eds.), *Proceedings of the 14th Arabic Conference of the Arab Union for Astronomy and Space Sciences* (Vol. 420, pp. 160–177). Springer Nature Singapore. https://doi.org/10.1007/978-981-96-3276-3_13

Holwerda, B. W., Trenti, M., Clarkson, W., Sahu, K., Bradley, L., Stiavelli, M., Pirzkal, N., Marchi, G. D., Andersen, M., Bouwens, R., Ryan, R., Vledder, I. V., & Vlugt, D. V. D. (2016). Erratum: "Milky Way red dwarfs in the BoRG survey; Galactic scale-height and the distribution of dwarf stars in WFC3 imaging" (2014, ApJ, 788, 77). The Astrophysical Journal, 825(1), 82. https://doi.org/10.3847/0004-637X/825/1/82

Huda, N., Zaki, A., Ali, A. K., Wahab, R. A., & Niri, M. A. (2014). Penentuan waktu solat Subuh menggunakan Rubu‘ Mujayyab di Malaysia: The determination of Subuh prayer time by using Rubu‘ Mujayyab in Malaysia. Jurnal Fiqh, *11*(11), 97–118.

Ilyas, M. (1984). A modern guide to astronomical calculations of Islamic calendar, times and qibla. In *A modern guide to astronomical calculations of Islamic calendar*. https://ui.adsabs.harvard.edu/abs/1984mgta.book.....I

Ilyas, M. (1997). *Astronomy of Islamic calendar* (1st ed.). A. S. Nordeen.

Izzuddin, A., Imroni, M. A., Imron, A., & Mahsun, M. (2022). Cultural myth of eclipse in a Central Javanese village: Between Islamic identity and local tradition. *HTS Teologiese Studies / Theological Studies*, *78*(4). https://doi.org/10.4102/hts.v78i4.7282

Junaidi, A. (2022). Syahadah Rukyatulhilal using astro digital imaging: From subjectivity to objectivity. *De Jure: Jurnal Hukum Dan Syar'iah*, *14*(1), 58–74. https://doi.org/10.18860/j-fsh.v14i1.15062

King, D. A. (1993). *Astronomy in the service of Islam*. Routledge.

Lairgi, Y. (2025). *When astronomy meets AI: Manazel for crescent visibility prediction in Morocco* (Version 1). arXiv. https://doi.org/10.48550/ARXIV.2503.21634

Loucif, S., Al-Rajab, M., Abu Zitar, R., & Rezk, M. (2024). Toward a globally lunar calendar: A machine learning-driven approach for crescent moon visibility prediction. *Journal of Big Data*, *11*(1), 114. https://doi.org/10.1186/s40537-024-00979-6

Meeus, J. (1991). *Astronomical algorithms*. Willmann-Bell.

Mehare, H. B., Anilkumar, J. P., & Usmani, N. A. (2023). The Python programming language. In M. "Sufian" Badar (Ed.), *A guide to applied machine learning for biologists* (pp. 27–60). Springer International Publishing. https://doi.org/10.1007/978-3-031-22206-1_2

Mohd Nawawi, M. S. A., Faid, M. S., Saadon, M. H. M., Wahab, R. A., & Ahmad, N. (2024). Hijri month determination in Southeast Asia: An illustration between religion, science, and cultural background. *Heliyon*, *10*(20). Scopus. https://doi.org/10.1016/j.heliyon.2024.e38668

Mustapha, A. S., Zainol, N. Z. N., Mohammad, C. A., Khalim, M. A., Muhammed, N. K. W., & Faid, M. S. (2024). Al-Fatani's perspectives on Islamic family law: Insights from Hidayah Al-Muta'allim Wa'Umdah Al-Muta'alim. *Journal of Islamic Thought and Civilization*, *14*(1), 247–265. https://doi.org/10.32350/jitc.141.15

Muztaba, R., Malasan, H. L., & Djamal, M. (2023). Deep learning for crescent detection and recognition: Implementation of Mask R-CNN to the observational Lunar dataset collected with the Robotic Lunar Telescope System. *Astronomy and Computing*, *45*, 100757. https://doi.org/10.1016/j.ascom.2023.100757

Niri, M. A., Jamaludin, M. H., Nawawi, M. S. A. M., Zaki, N. A., & Wahab, R. A. (2023). Astronomy development since antiquity to Islamic civilization from the perspective of Islamic historiography. *Journal of Al-Tamaddun*, *18*(1), 169–177. Scopus. https://doi.org/10.22452/JAT.vol18no1.14

Ozgur, C., Colliau, T., Rogers, G., & Hughes, Z. (2021). MatLab vs. Python vs. R. *Journal of Data Science*, *15*(3), 355–372. https://doi.org/10.6339/JDS.201707_15(3).0001

Rabbat, N. (1996). Al-Azhar mosque: An architectural chronicle of Cairo's history. *Muqarnas*, *13*, 45. https://doi.org/10.2307/1523251

Rasyid, M., Aseri, M., Izzuddin, A., Faid, M. S., Aseri, A. F., & Sukarni, S. (2024). Study of the Falakiyah Fatwas of the Indonesian Ulema council. *Journal of Law and Governance*, *7*(1), 45–62.

Rasyid, M., Aseri, M., Sukarni, Akh. Fauzi Aseri, & Faid, M. S. (2023). Science and its role in changes in Islamic legal thought (An analysis of changes in the Fatwa of the Indonesian Ulema council due to recent scientific findings). *Syariah: Jurnal Hukum Dan Pemikiran*, *23*(2), 120–137.

Razzak, A. (2024). A geophysical analysis through a Python program to predict sunrise, sunset, and prayer timings. *Jurnal Fizik Malaysia*, *45*(1), 10019–10027.

Rhodes, B. C. (2011). *PyEphem: Astronomical ephemeris for Python. Astrophysics Source Code Library, record* Ascl:1112.014.

Rubin, U. (2017). Morning and evening prayers in early Islam. In G. Hawting (Ed.), *The development of Islamic ritual* (1st ed., pp. 105–129). Routledge. https://doi.org/10.4324/9781315240312-9

Schumm, W. R. (2020). How accurately could early (622-900 C.E.) Muslims determine the direction of prayers (Qibla)? *Religions*, *11*(3), 102. https://doi.org/10.3390/rel11030102

Shariff, N. N. M., Faid, M. S., & Hamidi, Z. S. (2016). Islamic new moon software: Current status. In K. J. Kim & N. Joukov (Eds.), *Information science and applications (ICISA) 2016* (Vol. 376, pp. 1069–1079). Springer Singapore. https://doi.org/10.1007/978-981-10-0557-2_102

Shariff, N. N. M., Hamidi, Z. S., & Faid, M. S. (2017). The impact of light pollution on Islamic new moon (hilal) observation. *International Journal of Sustainable Lighting*, *19*(1), 10–14. https://doi.org/10.26607/ijsl.v19i1.61

Syazwan Faid, M., Mohd Nawawi, M. S. A., Saadon, M. H. M., & Ahmad, N. (2025). A review on modern lunar crescent visibility criterion. *Malaysian Journal of Science*, *44*(1), 101–120. https://doi.org/10.22452/mjs.vol44no1.12

Wahidi, A., Yasin, N., & Kadarisman, A. (2019). The beginning of Islamic months determination in Indonesia and Malaysia: Procedure and social condition. *ULUL ALBAB Jurnal Studi Islam*, *20*(2), 322–345. https://doi.org/10.18860/ua.v20i2.5913

Wahyuni, S., Samsuddin, S., & Hamzah, E. (2022). Qibla direction accuracy analysis based on astronomy (Google Earth), perspective of Islamic law. *Journal of Islam and Science*, *9*(1), 39–45. https://doi.org/10.24252/jis.v9i1.30111

Yildirim, F., Kadi, F., & Sahin, S. L. (2024). Developing a new interface for Qibla direction application based on MATLAB GUI. *Survey Review*, 1–15. https://doi.org/10.1080/00396265.2024.2411186

Yusuf, H. (2010). *Caesarean moon births: Calculations, moon sighting and the prophetic way*. Zaytuna Institute.

Index

THE PREMENSTRUAL SYNDROME

Proceedings of a workshop
held during the
Sixth International Congress of
Psychosomatic Obstetrics and Gynecology,
Berlin, September 1980

Edited by
Pieter A. van Keep and Wulf H. Utian
The editors were assisted by Pamela Freebody

MTP PRESS LIMITED
International Medical Publishers

Published by
MTP Press Limited
Falcon House
Lancaster, England

First published 1981

ISBN 0-85200-387-0

Printed and bound in Great Britain by
REDWOOD BURN LIMITED
Trowbridge and Esher

Contents

Principal participants

J. B. Day
Department of Obstetrics and Gynaecology,
St Thomas' Hospital,
London SE1 7EH,
United Kingdom

L. Dennerstein
Department of Psychiatry,
University of Melbourne,
Clinical Sciences Block,
c/o PO Royal Melbourne Hospital,
Victoria 3050,
Australia

A. A. Haspels
Academisch Ziekenhuis Utrecht,
Universiteitskliniek voor Obstetrie en Gynaecologie,
Catharijnesingel 101,
Utrecht,
The Netherlands

P. A. van Keep
International Health Foundation,
Rue de Rhône 40,
1204 Genève,
Switzerland

G. A. Sampson
'Queenswood' Psychiatric Unit,
Middlewood Hospital,
PO Box 134,
Sheffield S6 1TP,
United Kingdom

J. R. Strecker
Universität Ulm,
Frauenklinik,
Prittwitzstrasse 43,
7900 Ulm/Donau,
West Germany

W. H. Utian
Department of Obstetrics-Gynecology,
The Mount Sinai Hospital,
University Circle,
Cleveland, OH 44106,
USA

Preface

Despite a plethora of theories, premenstrual syndrome (PMS) has remained an enigma. There has persisted in the literature a constant conflict as to the existence of the syndrome, a question as to whether it is one syndrome or several, and a debate as to whether the origin is psychic, somatic, or both. Advances in endocrinology, specifically in radioligand assays, allowing for accurate hormone measurements, have precipitated a more scientific evaluation of PMS in recent years. Nonetheless, difficulties have persisted in accumulating well-documented data because of the protean nature of the syndrome. Indeed, even at this time, the question of what requires measurement during the follicular phase of the cycle and the premenstrual phase remains unresolved, and is difficult to place in perspective.

In view of the persisting conflict between the organic and the psychological schools of thought, we, the editors of this book, considered the Sixth International Congress of Psychosomatic Obstetrics and Gynecology to be an ideal venue for a workshop in which both parties could be encouraged to participate. Towards this end, the organizing committee of the International Society of Psychosomatic Obstetrics and Gynecology (ISPOG) was approached, and their response was extremely encouraging. In this respect, we wish to record our thanks to the Scientific Committee of ISOPG for allowing this workshop to be organized under their auspices, but totally under our direction. We, in turn, stand responsible for the format and content of the workshop.

The current proceedings represent the outcome of the workshop, and we express our thanks to the invited participants and to the general discussants.

It is our overall conclusion that PMS must be regarded as an endocrinopathy. It should be emphasized, however, despite the increasing tendency to regard the syndrome as a somatic disease, that there are certainly psycho–social–cultural aspects that contribute to symptom formation. This multi-factorial causation of symptoms accounts for the variable responses to different forms of therapy, including the marked response to placebos.

In presenting these proceedings we hope that the contents will act both as a source of information to clinicians and as a catalyst for further in-depth research into all aspects of PMS. We do believe that PMS as an endocrinopathy has 'come of age' and is accordingly deserving of a place in the future, not only at psychosomatic meetings, but also at scientific endocrine conventions. The surface of this problem has only been scratched; a considerable amount of in-depth research remains an urgent necessity.

Pieter A. van Keep
Wulf H. Utian

January 1981

1
Aetiology of premenstrual syndrome

J. B. DAY and R. W. TAYLOR

INTRODUCTION

Over the past 5 years a large number of papers have been written on the subject of premenstrual syndrome (PMS). These have covered the scientific basis of the syndrome, as well as treatment. A great deal of information has been collected, but no clear underlying cause of the syndrome has been found, probably because multiple factors are implicated.

Several theories as to the aetiology of PMS have been put forward, but it has often proved difficult to learn much from papers on this subject, firstly because many of the theories advanced are based on studies involving too few patients, and secondly because the conclusions reached by one author are seldom confirmed by another.

There is one area of knowledge, however, in which definite advances have been made, and that concerns the hormonal changes which take place during the menstrual cycle. It is likely that this is an area where still further important information will be obtained in the future.

In this paper we first briefly consider the psychogenic background to the PMS; we then discuss the hormonal factors which appear to be implicated in the aetiology of this syndrome, the way in which hormonal changes during the menstrual cycle may produce PMS symptoms, and finally the actions of the various types of treatment currently available. On this last point, ideally all treatments should be measured against other drugs and against placebo in controlled surroundings, but when treating PMS one must take care that this control does not produce its own set of inaccuracies. There is now good evidence that in PMS the placebo effect of treatment as such is very strong.

Psychogenic background

It has been stated that PMS is an evolutionary syndrome (Rosseinsky and Hall, 1974), the changing moods, with hostility and aggression during the premenstruum, increasing the male's ardour while keeping him at a distance until the fertile period approaches – this fertile period coinciding with a peak of human female sexuality at mid-cycle (Adams *et al.*, 1978). This, in 1980, is probably an unrealistic theory. It is true, however, that for some women menstruation becomes an unnecessary intrusion into an already difficult life. Premenstrual tension, with its depression and irritability, may be a reflection of this. Paige (1971), confirming the work of Grant and Pryse-Davies (1968), found that despite the presence of menstruation women on combined or sequential oral contraceptives do not experience these cyclic mood changes. It seems, however, that today some women taking these preparations do experience PMS, a fact which may well be related to the lowering in recent years of the hormonal content of many of the contraceptive pills available.

Berry and McGuire (1972), using a questionnaire, related sexual role to cyclical symptoms and found no evidence that PMS symptoms were associated with psychosocial or psychosexual aspects of women's lives. Menstruating women, however, appear to have greater mood swings during their cycle than those who have had a hysterectomy (Beumont *et al.*, 1975).

If the complaint by women of their normal physiology is considered an abnormal response, then these women do indeed have a neurotic tendency, but Pennington (1957) found that 95% of women have premenstrual symptoms, thereby suggesting that PMS is intrinsic and physiological. The problem of relating PMS to neuroticism is well described by Clare (1977), who pointed out that the symptoms used to score neuroticism overlap heavily with those used to score PMS. Coppen and Kessel (1963) found a positive correlation between neuroticism and severe PMS in 500 patients randomly sampled; this may be a valid point, but we are no nearer answering the question of aetiology. It is also interesting to note that PMS occurs in the mentally retarded in whom frustration and emotional disturbances are rarely seen (Takayama, 1972). The findings of Kantero and Widholm (1971) suggest either a genetic element to the condition or a psychosomatic induction of the syndrome, since 63% of daughters of unaffected mothers are symptom-free, whereas 70% of the daughters of affected mothers are themselves PMS sufferers.

The psychogenic influences surrounding a woman during her menstrual cycle may adversely affect her psychologically, thereby increasing the magnitude of any existing problems. The occurrence of menstruation, the awareness of ovulation, and then premenstrual symptoms, may form a potential cycle of stress for any woman. The stress produced may well increase anxiety and depression during the premenstruum.

PMS, to some, is a major disturbance but one which is well-accepted. To relate 'neuroticism' to such people, when their problem is brought about by a biochemical change, seems unhelpful when discussing aetiology.

In summary, PMS for any individual will have a considerable input from the woman's personality, which in turn will be affected by her surroundings. The individual's stress may further exaggerate an inherent biochemical change.

Ovarian steroids

Frank in 1931 focused on an 'oestrogen excess' as a causative factor of PMS, and Israel (1938) thought it not an excess of oestrogen but a 'lack of progesterone antagonism'. Hormonal measurements in those times were all indirect, and Israel demonstrated an abnormal secretory phase by endometrial biopsy. He could not, however, understand why PMS was not seen in anovular cycles.

Other workers followed using indirect assessment of hormone function in the luteal phase, and all basically supported the idea that the progesterone to oestrogen ratio is low in PMS patients (Greene and Dalton, 1953; Rees, 1953; Morton, 1950).

The direct evidence of the phase of ovarian steroids comes from work by Bäckström and Carstensen (1974) and Munday (1977) who performed plasma estimations of progesterone and

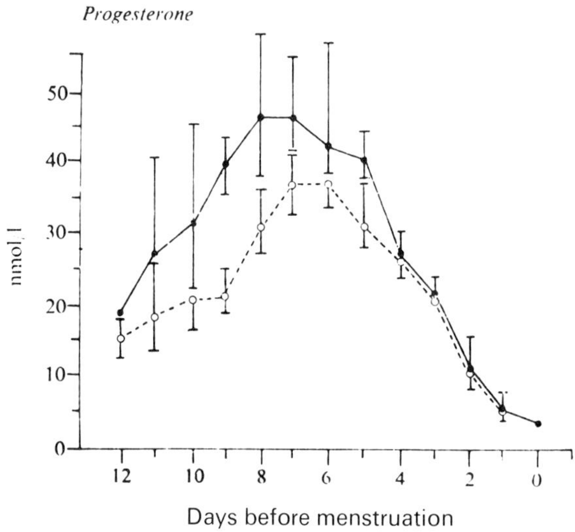

Figure 1 Logarithmic mean (± SEM) of plasma progesterone levels prior to menstruation in 20 women with premenstrual syndrome (○----○) and in ten control cycles (●——●). (Munday 1979. Reproduced with permission)

oestrogen or oestradiol, and it is these findings which form the basis of our present understanding. A trend has been found by these workers, both initially and more recently, but it should be pointed out that the daily fluctuations found by others have not always confirmed the findings of these two groups.

Bäckström and Carstensen (1974) compared daily steroids from eight women with no PMS with those of ten women with premenstrual anxiety and irritability. The progesterone levels in days 1–3 prior to menstruation were the same for both groups, but in days 4–10 premenstrually both oestrogen and progesterone were low in a small group with PMS. The oestrogen values rose to a peak above controls 4 days premenstrually, while progesterones continued to be low. There were no differences in LH values or in levels of the sex hormone binding globulin. FSH levels were higher in the PMS group 6–9 days premenstrually, after which they fell to the levels in control subjects.

Munday and her co-workers at St Thomas' Hospital in London found there to be lower progesterone levels in the PMS

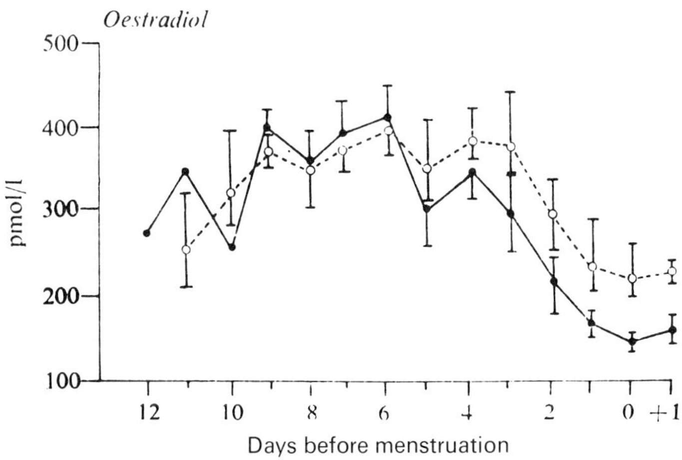

Figure 2 Logarithmic mean (±SEM) of plasma oestradiol levels prior to menstruation in 20 women with premenstrual syndrome (○– – –○) and in ten control cycles (●——●). (Munday 1979. Reproduced with permission)

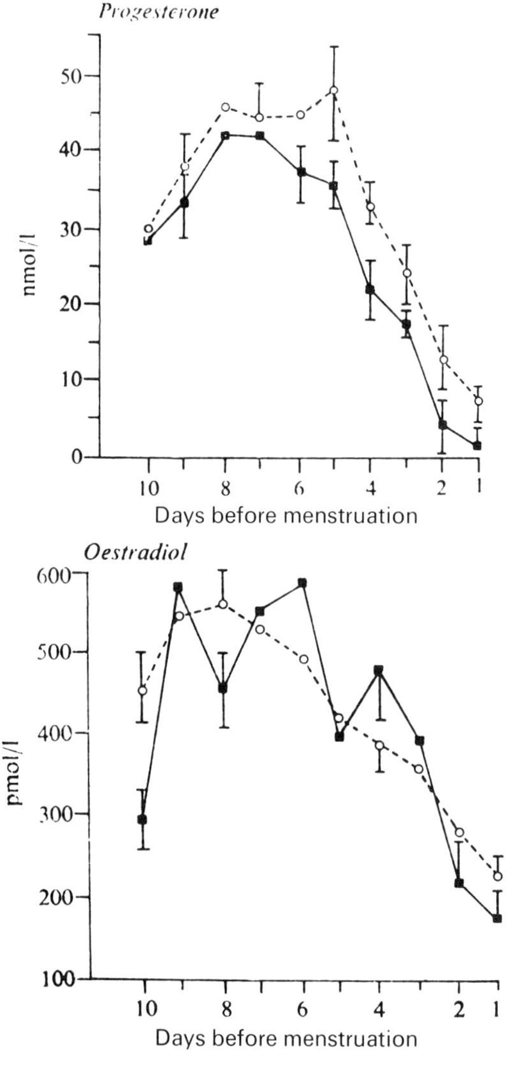

Figure 3 Daily mean (± SEM) plasma progesterone and oestradiol levels in ten control patients (■——■) compared with mean values for controls (○- - - -○). (Munday, 1979. Reproduced with permission. The control data are those of Bäckström, who has kindly given permission for their reproduction here)

group at the mid-luteal phase from days 5–10 premenstrually (Figure 1). Munday concluded that after ovulation progesterone seemed impaired in the PMS group. The oestradiol levels were similar to those of the controls but were slightly higher in PMS patients after the mid-luteal peak of progesterone. A significant difference was shown from 4 days before menstruation until 1 day after it (Figure 2). The progesterone to oestradiol ratio was also calculated and found to be significantly higher in the controls for the last 8 days of the cycle. Munday's controls compared to those of Bäckström for plasma progesterone and oestradiol are shown in Figure 3. Smith (1975) found lowered progesterone levels, but no difference in plasma oestrogen.

It is only now that the significance of these first plasma results becomes apparent. They must be confirmed with as many as two samples per day in the follicular phase as well as in the luteal phase in patients with well-defined PMS and in controls.

Aldosterone and the renin angiotensin system

There is no evidence that aldosterone is a prime factor in the aetiology of PMS, although a premenstrual rise in plasma aldosterone in PMS patients was shown by Brush (1977) and by O'Brien *et al.* (1979). This rise, however, was no greater than in controls. The rise may be due to the progesterone levels of the luteal phase producing a natriuresis, followed by increased secretion of aldosterone to maintain homeostasis. The other possibility is that aldosterone metabolism may directly affect brain function. There is no evidence of this for aldosterone, but angiotensin appears to alter behaviour, central neurotransmitters and autonomic function in animals (Janowsky *et al.*, 1973). It could, however, be postulated that an altered progesterone to aldosterone ratio may alter the stability of the RAA system and hence lead to symptoms of PMS. It is also interesting to note that several patients of O'Brien (1979), although complaining of bloatedness, showed no weight gain, suggesting that there are changes within the extra- and intra-vascular compartments. Aldosterone is a powerful antinatriuretic hormone, but so are oestrogens and antidiuretic hormone, and, to a lesser extent, prolactin and cortisol.

Prolactin

The place of prolactin is interesting, partly because the results available are conflicting. It is clear, however, that prolactin release is pulsatile (McNeilly and Chard, 1974), and that prolactin levels are not, therefore, constant.

Horrobin (1973) suggested that prolactin could be responsible both for the mental and for the physical symptoms of PMS. He was able to show that prolactin retains sodium and potassium.

Direct evidence in the form of serum measurements comes from Halbreich *et al.* (1976): four measurements made during the last 3 weeks of the cycle in 28 patients with PMS and in 21 controls showed that the PMS patients had a higher mean prolactin level and that their individual increases were also greater.

The use of bromocriptine has added further information on the place of prolactin. In controlled cross-over trials, Benedek-Jaszmann and Hearn-Sturtevant (1976) improved breast symptoms, oedema, weight gain and mood in patients attending an infertility clinic. Ghose and Coppen (1977), however, did not improve PMS patients by using bromocriptine; Andersen *et al.* (1977) improved mastodynia but none of their patients' other PMS symptoms. Five of the Andersen *et al.* patients had transient high levels of prolactin. Andersch *et al.* (1978) studied 20 PMS patients and 20 controls and found a luteal increase in prolactin; they also found that treatment with bromocriptine improved the PMS symptoms of breast discomfort and irritability.

The potential effect of prolactin on the length of the menstrual cycle should also be considered. Pronounced hyperprolactinaemia leads to amenorrhoea, but slightly raised levels produce a shortened luteal phase or are associated with this (Corenblum *et al.*, 1976; Benedek-Jaszmann and Hearn-Sturtevant, 1976). Two forms of inadequate luteal phase are described: a short luteal phase lasting less than 10 days with lowered steroid secretion (Strott *et al.*, 1970; Sherman and Korenman, 1974), and a defective luteal phase of normal length but with lowered steroid secretion (Pearce *et al.*, 1971; Dodson *et al.*, 1975). The aetiology of these defects is not known, but prolactin hypersecretion was implicated by Lenton *et al.* (1977) who noted slightly raised prolactin levels in the early follicular phase in women with a suspected, though not proven, inadequate luteal phase. Workers at St Thomas' Hospital found a raised prolactin level in association with a low progesterone level in PMS patients compared to controls (Brush, 1979) (Figure 4).

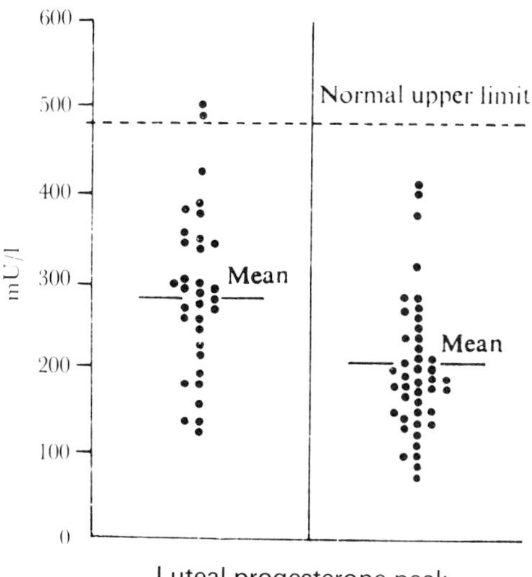

Figure 4 Comparison of serum prolactin levels in 33 premenstrual syndrome subjects with a 'low' progesterone curve (left panel) and in 42 subjects with a 'normal' progesterone curve

Precise interactions of prolactin with other hormones during the menstrual cycle are not known. Prolactin is known to inhibit the positive feedback of oestradiol on the LH peak at mid-cycle (Aono *et al.*, 1976; Glass *et al.*, 1975); the LHRH response from the pituitary in patients with hyperprolactinaemia is unaffected (Mortimer *et al.*, 1973; Zarate *et al.*, 1973), implying that the positive feedback of oestradiol to the hypothalamus is blocked in hyperprolactinaemia. Prolactin might also have a direct effect on the ovary (McNatty *et al.*, 1974).

The prolactin effect in the production of PMS may be via the lowering of progesterone and oestradiol levels during the luteal phase, or it may act directly on mineral and water metabolism. It may act directly on the breast, but not all hyperprolactinaemia patients develop galactorrhoea (Seppälä *et al.*, 1977).

Other theories on the aetiology of PMS

Hyperinsulinism in the premenstrual phase was suggested as a cause of PMS by Billig and Spaulding (1947). We now know that hypoglycaemia would cause a transient rise in prolactin.

An allergic response to steroids was suggested by Rogers (1962), who claimed successful treatment with minute quantities (0.1–1.0 μg) of pregnanediol injected subcutaneously.

Several vitamins have been thought to be implicated. The first mentioned was vitamin B (Biskind, 1943). More recently pyridoxine deficiency (vitamin B_6) has been suggested as a potential cause of the syndrome (Brush, 1977); in the form of pyridoxal-5-phosphate, vitamin B_6 is a co-factor in a number of the body's enzyme reactions. Particularly important is its place as a co-enzyme in the production of dopamine from tyrosine and of serotonin from tryptophan. At present there is no information regarding pyridoxal phosphate blood levels in PMS patients, though an assay has been described by Bhagavan *et al.* (1976).

A relative deficiency of vitamin B_6 has been suggested as a cause of depression in patients taking the combined oral contraceptive pill; this has been successfully treated with pyridoxine (Adams *et al.*, 1973, following an initial observation by Rose, 1969).

Treatment of PMS patients with pyridoxine was disappointing for Stokes and Mendels (1972), but Kerr (1977) and Day (1979) found it helpful to some extent. More controlled studies are required in this connection, together with assays of pyridoxal phosphate, though these may prove disappointing.

AN AETIOLOGICAL MODEL (Figure 5)

From the data available at present there is a definite suggestion that there are alterations in the progesterone and oestradiol levels in some PMS patients. This has the effect of producing a lower progesterone to oestradiol ratio during the last 7–10 days of the menstrual cycle, the progesterone levels being low in the early- and mid-luteal phase, while oestradiol levels are normal, the oestradiol levels then rising significantly 4 days premenstrually. Added to this is the evidence from Bäckström *et al.* (1976) that FSH levels are higher in the mid-luteal phase.

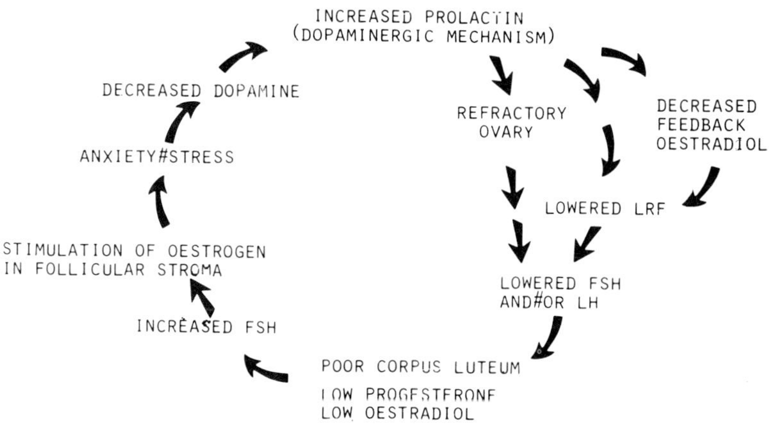

Figure 5 Aetiological model

Finally, the variability of prolactin must be included, with a tendency for raised levels in patients with PMS. A raised prolactin level early in the cycle, even if transient, due to stress or a hypothalamic factor, may lead to impaired follicular development and abnormal secretion of oestrogen and progesterone. The LH peak may not be affected, but a defective luteal phase may still follow. The abnormal follicle and altered level of oestrogen and progesterone would then allow a positive feedback to the pituitary and/or hypothalamus, and a luteal surge of FSH. In response to FSH the oestradiol levels may then start to rise towards the end of the luteal phase; a second LH surge does not occur, however, because of the presence of progesterone. Once established, the asynchronous cycle may continue with increasing stress factor induced by the cycle itself. This asynchronous potential of follicular development can be demonstrated in women by the number of follicles in the human ovary at different stages of development (Fowler *et al.*, 1977).

Mechanisms for an altered progesterone to oestradiol ratio are via:

(1) an alteration of the steroid pathway within the corpus luteum, or because of an absolute deficiency or imbalance of granulosa and/or of thecal cells;
(2) the effect of gonadotrophin producing an effect on the cells of the corpus luteum or on other ovarian components;
(3) a direct effect of prolactin on steroid steroidogenesis.

THE PRODUCTION OF SYMPTOMS

All the hormones mentioned so far have powerful effects on the movement of minerals and water within the body. What is difficult to know is whether it is a central change that produces the peripheral changes or whether it is a local action which produces them.

There is now good evidence that the gonadal steroids oestradiol and progesterone accumulate in the central nervous system, and that they affect brain metabolism and neural function. When injected into rats, ^{3}H progesterone is taken up very quickly by the cortex, sub-cortex, brain stem and cerebellum, but the radioactivity is lost more quickly from these areas than from the hypothalamus.

The general impression is that progesterone is tranquillizing and hypnotic, and that oestrogen tends to increase nervous activity; a lowering of the progesterone to oestradiol ratio could, therefore, be expected to lead to an increase in tension and irritability.

The contradictory evidence is that the progesterone to oestrogen ratio is at its lowest in the follicular phase, with no equivalent pre-ovulation syndrome. In view of this, individual levels are probably not important, but thresholds of action of any particular hormone are. It may be envisaged that hormone action will only produce an effect after interaction with other chemicals or with neurotransmitters. The pharmacological effects produced may be either to modify the response of a hormone on a receptor site or to actually alter the binding site availability.

The biogenic amines are probably mediating factors where hormones do not act quickly; the two main groups are the indoleamines, such as serotonin, which is responsible for serotonergic transmitter systems, and the catecholamines, such as dopamine and noradrenaline, which are involved, respectively, in cholinergic and noradrenergic systems.

In the future the link between these amines and the regulation of hypothalamic releasing factors will be an important area of research.

Regarding the somatic symptoms, these are best divided into (a) breast symptoms and (b) fluid retention symptoms. The breast symptoms probably occur as a direct result of an end-organ hormone imbalance, with prolactin featuring strongly.

The fluid retention symptoms, on the other hand, are probably a general reflection of an altered fluid transference between cells, and here the changes of progesterone to aldosterone, progesterone to oestradiol, and/or progesterone to prolactin ratios are important.

THE MODE OF ACTION OF TREATMENTS

Progesterone and progestogens

Firstly, some progestogens promote symptoms of their own, a particular example being norethisterone with its powerful water-retaining characteristics. The efficacy of natural progesterone is difficult to explain, as it is unlikely to be related to the making up of a deficiency and thereby to improving, for example, the progesterone to oestradiol ratio.

We have had good experience with dydrogesterone, an orally administered progestogen which is well-tolerated by patients. We have found it particularly beneficial in connection with depression, anxiety and irritability, improvements which probably occur as a result of a central effect on the areas of the cortex producing changes of mood. We have found it less useful for the physical symptoms of PMS. Some of our recent studies have suggested that the progesterone level during the luteal phase is actually reduced in patients taking dydrogesterone. Other studies, however, have not found this to be so, a fact which further exemplifies how difficult it is to interpret isolated hormone levels in this syndrome.

Aldosterone inhibitor

The diuretic spironolactone should, through its anti-aldosterone action, release body fluid. This it does, and patients suffering premenstrual oedema find it very useful. The oedema does, however, return as soon as the administration of spironolactone is stopped, suggesting that its action is almost entirely within the kidney. A potential benefit of spironolactone is its progesterone-raising effect, probably via the adrenals. This may produce an overall improvement of the syndrome. A dydrogesterone/spironolactone preparation started just prior to mid-cycle would be an interesting thought if the syndrome is severe. It may be, however, that the two compounds would competitively block each other.

Dopamine

Bromocriptine has had disappointing results except for its effect on mastodynia and mastalgia, a local action. Its lack of effect on other symptoms makes one inclined to think that the extra dopamine provided is unnecessary; it does not, for instance, 'raise' neurotransmission, improve mood, etc. If it could be shown that a patient had a dopamine shortage, bromocriptine may be effective for the treatment of the whole syndrome.

Antigonadotrophin

Danazol is a drug which is still under-evaluated as far as PMS is concerned. It produces a dramatic response in some patients, but it is difficult to predict in which ones. Small doses of danazol may well, through the changes it produces in the hypothalamus, be able to re-programme the timing of FSH release, the response to oestradiol, the LH surge, etc., and so actually stop the symptoms of PMS occurring. Clinical trials with this substance are needed.

Pyridoxine

There is no evidence that vitamin B_6 deficiency occurs as a result of dietary deficiency in northern European countries. This being so, vitamin B_6 replacement is illogical, unless it can be shown that it centrally blocks other symptom-producing transmitter substances. At this stage it is our feeling that the effects of pyridoxine in patients suffering PMS are those of a placebo.

Monamine oxidase inhibitors

These have been widely used with good effects in patients suffering depressive illnesses, but there are only a small number of PMS sufferers who would benefit from them. It is difficult to know how they produce their beneficial effects; it could be by permitting the accumulation of selected amines, or by protection from the metabolism of certain amines, or it may be that they act by maintaining amine levels in an area of homeostasis, such as the hypothalamus.

Placebo effect

Finally, it must be said that the placebo aspect is certainly one of the most effective parts of treatment, and although severe symptoms may not be improved dramatically there will often be some amelioration. The mere fact that a patient discusses the problem with her doctor, and receives sympathetic support, may well cause the stress factor to be palliated, and physiology at the hypothalamic level to return to normal.

References

Adams, D. B., Gold, A. R. and Burt, A. D. (1978). Rise in female initiated sexual activity at ovulation and its suppression by oral contraceptives. *New Engl. J. Med.*, **229**, 1145

Adams, P. W., Rose, D. P., Folkard, J., Wynn, V., Seed, M. and Strong, R. (1973). Effect of pyridoxine hydrochloride (vitamin B_6) upon depression associated with oral contraception. *Lancet*, **1**, 897

Andersch, B., Hahn, L., Wendestam, C., Öhman, R. and Abrahamsson, L. (1978). Treatment of premenstrual tension syndrome with bromocriptine. *Acta Endocrinol.*, **88** (Suppl. 216), 165

Andersen, A. N., Larsen, J. F., Steenstrup, O. R., Svendstrup, B. and Nielsen, J. (1977). Effect of bromocriptine on the premenstrual syndrome. A double-blind clinical trial. *Br. J. Obstet. Gynaecol.*, **84**, 370

Aono, T., Miyake, A., Shioji, T., Kinugasa, T., Onishi, T. and Kurachi, K. (1976). Impaired LH release following exogenous estrogen administration in patients with amenorrhea galactorrhea syndrome. *J. Clin. Endocrinol.*, **42**, 696

Bäckström, T. and Carstensen, H. (1974). Estrogen and progesterone in plasma in relation to premenstrual tension. *J. Steroid Biochem.*, **5**, 257

Bäckström, T., Wide, L., Södergård, R. and Carstensen, H. (1976). FSH, LH, Te-BG-capacity, estrogen and progesterone in women with premenstrual tension during the luteal phase. *J. Steroid Biochem.*, **7**, 473

Benedek-Jaszmann, L. J. and Hearn-Sturtevant, M. D. (1976). Premenstrual tension and functional infertility. *Lancet*, **1**, 1095

Berry, C. and McGuire, F. L. (1972). Menstrual distress and acceptance of sexual role. *Am. J. Obstet. Gynecol.*, **114**, 83

Beumont, P. J. V., Richards, D. H. and Gelder, M. G. (1975). A study of minor psychiatric and physical symptoms during the menstrual cycle. *Br. J. Psychiatry*, **126**, 431

Bhagavan, H. N., Koogler, J. M. and Coursin, D. B. (1976). Enzymatic microassay of pyridoxal-5′-phosphate using L-tyrosine apo-decarboxylase and L-(1-^{14}C) tyrosine. *Int. J. Vitam. Nutr. Res.*, **46**, 160

Billig, H. E. Jr. and Spaulding, C. A. Jr. (1947). Hyperinsulinism of menses. *Ind. Med.*, **16**, 336

Biskind, M. S. (1943). Nutritional deficiency in the etiology of menorrhagia, metrorrhagia, cystic mastitis and premenstrual tension; treatment with vitamin B complex. *J. Clin. Endocrinol.*, **3**, 227

Brush, M. G. (1977). The possible mechanisms causing the premenstrual tension syndrome. *Curr. Med. Res. Opin.*, **4** (Suppl. 4), 9

Brush, M. G. (1979). Endocrine and other biochemical factors in the aetiology of the premenstrual syndrome. *Curr. Med. Res. Opin.*, **6** (Suppl. 5), 19

Clare, A. W. (1977). Psychological profiles of women complaining of premenstrual symptoms. *Curr. Med. Res. Opin.*, **4** (Suppl. 4), 23

Coppen, A. and Kessel, N. (1963). Menstruation and personality. *Br. J. Psychiatry*, **109**, 711

Corenblum, B., Pairaudeau, N. and Shewchuk, A. B. (1976). Prolactin hypersecretion and short luteal phase defects. *Obstet. Gynecol.*, **47**, 486

Day, J. B. (1979). Clinical trials in the premenstrual syndrome. *Curr. Med. Res. Opin.*, **6** (Suppl. 6), 40

Dodson, K. S., MacNaughton, M. C. and Coutts, J. R. T. (1975). Infertility in women with apparently ovulatory cycles. I. Comparison of their plasma sex steroid and gonadotrophin profiles with those in the normal cycle. *Br. J. Obstet. Gynaecol.*, **82**, 615

Fowler, R. E., Chan, S. T., Walters, D. E., Edwards, R. G. and Steptoe, P. C. (1977). Steroidogenesis in human follicles approaching ovulation as judged from assays of follicular fluid. *J. Endocrinol.*, **72** (3), 259

Frank, R. T. (1931). The hormonal causes of premenstrual tension. *Arch. Neurol. Psychiatry (Chic.)*, **26**, 1053

Ghose, K. and Coppen, A. (1977). Bromocriptine and premenstrual syndrome: controlled study. *Br. Med. J.*, **1**, 147

Glass, M. R., Shaw, R. W., Butt, W. R., Logan Edwards, R. and London, D. R. (1975). An abnormality of oestrogen feedback in amenorrhoea–

galactorrhoea. *Br. Med. J.*, **3**, 274
Grant, E. C. G. and Pryse-Davies, J. (1968). Effect of oral contraceptives on depressive mood changes and on endometrial monamine oxidase and phosphatases. *Br. Med. J.*, **3**, 777
Greene, R. and Dalton, K. (1953). The premenstrual syndrome. *Br. Med. J.*, **1**, 1007
Halbreich, U., Ben-David, M., Assael, M. and Bornstein, R. (1976). Serum-prolactin in women with premenstrual syndrome. *Lancet*, **2**, 654
Horrobin, D. F. (1973). *Prolactin: Physiology and Clinical Significance* (Lancaster: MTP)
Israel, S. L. (1938). Premenstrual tension. *J. Am. Med. Assoc.*, **110**, 1721
Janowsky, D. S., Berens, S. C. and Davis, J. M. (1973). Correlations between mood, weight and electrolytes during the menstrual cycle: a renin–angiotensin–aldosterone hypothesis of premenstrual tension. *Psychosom. Med.*, **35**, 143
Kantero, R. L. and Widholm, O. (1971). Correlations of menstrual traits between adolescent girls and their mothers. *Acta Obstet. Gynecol. Scand.*, **14** (Suppl. 14), 30
Kerr, G. D. (1977). The management of the premenstrual syndrome. *Curr. Med. Res. Opin.*, **4** (Suppl. 4), 29
Lenton, E. A., Sobowale, O. S. and Cooke, I. D. (1977). Prolactin concentrations in ovulatory but infertile women: treatment with bromocriptine. *Br. Med. J.*, **2**, 1179
McNatty, K. P., Sawers, R. S. and McNeilly, A. S. (1974). A possible role for prolactin in control of steroid production by the human Graafian follicle. *Nature*, **250**, 653
McNeilly, A. S. and Chard, T. (1974). Circulating levels of prolactin during the menstrual cycle. *Clin. Endocrinol.*, **3**, 105
Mortimer, C. H., Besser, G. M., McNeilly, A. S., Marshall, J. C., Harsoulis, P., Tunbridge, W. M. G., Gomez-Pan, A. and Hall, R. (1973). Luteinizing hormone and follicle stimulating hormone tests in patients with hypothalamic–pituitary–gonadal dysfunction. *Br. Med. J.*, **4**, 73
Morton, J. H. (1950). Premenstrual tension. *Am. J. Obstet. Gynecol.*, **60**, 343
Munday, M. (1977). Hormone levels in severe premenstrual tension. *Curr. Med. Res. Opin.*, **4** (Suppl. 4), 16
Munday, M. R. (1979). Endocrine studies in the premenstrual syndrome. Ph.D. Thesis, University of London, UK
O'Brien, P. M. S., Craven, D., Selby, C. and Symonds, E. M. (1979). Treatment of premenstrual syndrome by spironolactone. *Br. J. Obstet. Gynaecol.*, **86**, 142
Paige, K. E. (1971). Effects of oral contraceptives on affective fluctuations associated with the menstrual cycle. *Psychosom. Med.*, **33**, 515
Pearce, M. A., Fahmy, D., Morgan, C., Evans, C., Groom, G., Boyns, A. R. and Cooke, I. D. (1971). Hormonal changes in patients with defective luteal phases. *J. Endocrinol.*, **51**, 24

Pennington, V. M. (1957). Meprobromate (Miltown) in premenstrual tension. *J. Am. Med. Assoc.*, **164**, 638

Rees, L. (1953). The premenstrual tension syndrome and its treatment. *Br. Med. J.*, **1**, 1014

Rogers, W. C. (1962). The role of endocrine allergy in the production of premenstrual tension. *West. J. Surg. Obstet. Gynaecol.*, **70**, 100

Rose, D. P. (1969). Oral contraceptives and depression. *Lancet*, **2**, 321

Rosseinsky, D. R. and Hall, P. G. (1974). An evolutionary theory of premenstrual tension. *Lancet*, **2**, 1024

Seppälä, M., Lehtovirta, P. and Ranta, T. (1977). Discordant patterns of hyperprolactinaemia and galactorrhoea in secondary amenorrhoea. *Acta Endocrinol. (Kbh)*, **86**, 457

Sherman, B. M. and Korenman, S. G. (1974). Measurement of plasma LH, FSH, oestradiol and progesterone in disorders of the human menstrual cycle: the short luteal phase. *J. Clin. Endocrinol.*, **38**, 89

Smith, S. L. (1975). Mood and the menstrual cycle. In Sacher, E. (ed.) *Topics in Psychoendocrinology*, pp. 19–58 (New York: Grune and Stratton)

Stokes, J. and Mendels, J. (1972). Pyridoxine and premenstrual tension. *Lancet*, **1**, 1177

Strott, C. A., Cargille, C. M., Ross, G. T. and Lipsett, M. B. (1970). The short luteal phase. *J. Clin. Endocrinol.*, **30**, 246

Takayama, T. (1972). A clinical study of premenstrual syndrome. In Morris, N. (ed.) *Psychosomatic Medicine in Obstetrics and Gynaecology*, p. 596 (Basel: Karger)

Taylor, R. W. and James, C. E. (1979). The clinician's view of patients with premenstrual syndrome. *Curr. Med. Res. Opin.*, **6** (Suppl. 5), 46

Zarate, A., Jacobs, L. S., Canales, E. S., Schally, A. V., de la Cruz, A., Soria, J. and Daughaday, W. H. (1973). Functional evaluation of pituitary reserve in patients with amenorrhoea–galactorrhoea syndrome utilizing luteinizing-hormone–releasing hormone. *J. Clin. Endocrinol.*, **37**, 855

2
The premenstrual syndrome – an epidemiological and statistical exercise

P. A. VAN KEEP and P. LEHERT

INTRODUCTION

Premenstrual syndrome is the collective name given to the problems which occur during the 2-week period before menstruation, and which are considered to be a prelude to menstruation. The list of symptoms regarded as being part of this syndrome is long and varied.

It is somewhat strange when one considers how many women appear to suffer from this phenomenon that more precise knowledge as to the extent of the problem is not available. Most of the few reports which have appeared on the subject have been in connection with studies conducted in very selected groups, and the figures relating to the incidence of the problem have varied considerably (Table 1).

It is strange too that the premenstrual syndrome, first described by Frank in 1931, is still today so vaguely defined. This fact also makes the various studies and reports difficult to compare. Premenstrual syndrome, as it is 'defined' at present, seems to differ considerably from one woman to another, which sometimes makes it difficult for a woman and for her doctor to recognize a particular problem as being part of this syndrome (Dalton, 1980). The matter is further complicated by the fact that the duration of the symptoms and their severity also appear to vary enormously. Many women experience their premenstrual symptoms for 2 or 3 days, but some regularly experience them for a week or longer, and whilst some women see these symptoms as little more than a quickly passing sign that menstruation is on its way, others are incapacitated to the extent that it is impossible for them to continue with their normal daily routines.

Table 1 Studies reporting the incidence of premenstrual syndrome (PMS)

Authors	*Year*	*Group*	*Number*	*% with PMS*
Bickers and Wood	1951	Factory workers (USA)	± 1500	36
Morton *et al.*	1953	Female prisoners (USA)	249	51
Kessel and Coppen	1963	GP practices (UK)	500	25
Sutherland and Stewart	1965	Hospital personnel and students (UK)	150	33
Herschberg	1966	Gynaecology patients (France)	5000	32
Wetzel *et al.*	1975	Students (USA)	589	40
Clare	1977	GP practices (UK)	521	75
van Keep and Haspels	1979	National probability sample (The Netherlands)	1493	47

It is important to obtain more insight into the frequency of premenstrual syndrome in the female population as a whole, because only then shall we understand how far this is a health problem. Epidemiological studies may also provide insight into the type of woman who is particularly prone to premenstrual syndrome, and may therefore help the clinician to recognize a patient.

It was against this background that the study reported in the present paper was undertaken. The statistical exercise was conducted on data obtained by the International Health Foundation during a study in France in 1978–9 (International Health Foundation, 1979). A few of the data from the survey are repeated here in order to explain the background to the exercise.

THE STUDY

The study was a postal one conducted in France with the help of Sofres Médical, the medical branch of a national opinion poll company, in December 1978 and January 1979. It involved 2501 women between the ages of 15 and 50 from all parts of the country. In order to obtain the desired sample size it was necessary to send out 3600 questionnaires. The return rate, therefore, was 69% (a rate which would possibly have been higher had the survey been conducted at a less busy time of year).

The questionnaire contained 39 questions, and great emphasis was laid on the fact that the symptoms being discussed were those occurring during the 2 weeks *before* menstruation. Twenty-five possible symptoms were listed in alphabetical order. They were:

Back-ache, kidney ache
Clumsiness, proneness to accidents
Cold sweats
Difficulty in concentrating and in remembering things
Dizziness, a feeling of being not quite well
Eye or sight problems
General 'out-of-sorts' feeling
Headaches
Hot flushes
Indecision, inefficiency
Insomnia
Moodiness, irritability, aggressiveness

Muscle stiffness
Ovarian pains
Painful, heavy legs
Painful, swollen, tender breasts
Pains in the pelvic area
Sickness, nausea
Skin problems (irritations, blotchiness)
Sleepiness, lethargy
Stomach ache
Tearfulness, depression
Tenseness, general uneasiness
Tiredness, general weakness
Weight-gain, puffiness

A space was also left in which the respondents could add other symptoms which they felt to be part of the premenstrual syndrome. Thirty-five women did so, but none of the symptoms added featured more than two or three times, so it is felt that the above list can probably be regarded as a fairly comprehensive one, at least as far as the lay public is concerned.

Of the 2501 women who completed the questionnaire, 943 (38%) said that they experienced premenstrual pains or symptoms every month. A further 684 (27%) said that they experienced them from time to time, and 290 (12%) said that they experienced them occasionally. Thus, 77% of the 2501 women taking part in this survey reported that they experienced premenstrual symptoms. In reply to a later question, only 354 (14%) of the 2501 respondents said that they had never experienced such symptoms.

Two factors are probably involved in the moving of a premenstrual symptom into the category of a premenstrual complaint or problem: (a) the time it lasts, and (b) its severity. In this study the first factor was covered simply by asking the respondents to indicate the number of days on which they usually had the symptom; the second was covered by asking them to indicate if each symptom troubled them a little, moderately or very much. The most frequently experienced symptoms, together with the number and percentage of women usually experiencing each of the problems for 7 days or more, and the number and percentage of women troubled 'very much' by each problem, are given in Table 2. The incidence of the less extensively experienced symptoms is summarized in the footnote.

Table 2 Number of women experiencing premenstrual symptoms, those usually experiencing such problems for 7 or more days, and those being troubled by them 'very much'. The percentage calculations are based on all 2501 respondents

	Women experiencing the premenstrual symptom		*Women usually experiencing the symptom for 7+ days*		*Women being troubled 'very much' by the symptom*	
	No.	%	*No.*	%	*No.*	%
Moodiness, irritability, aggressiveness	926	37	180	7	220	9
Painful, swollen, tender breasts	829	33	291	12	147	6
Back-ache, kidney ache	773	31	91	4	193	8
Pains in the pelvic area	771	31	77	3	206	8
Ovarian pains	643	26	73	3	176	7
Tiredness, general weakness	630	25	137	5	124	5
Weight-gain, puffiness	594	24	176	7	131	5
Painful, heavy legs	470	19	93	4	104	4
Tearfulness, depression	461	18	78	4	115	5
Tenseness, general uneasiness	451	18	111	4	116	7
Headaches	428	17	34	1	166	7

The following symptoms were also mentioned: muscle stiffness (13%), general 'out-of-sorts' feeling (12%), skin problems (irritations, blotchiness) (11%), hot flushes (11%), insomnia (8%), sickness, nausea (7%), difficulty in concentrating and in remembering things (5%), sleepiness, lethargy (5%), dizziness, a feeling of being not quite well (6%), stomach ache (3%), clumsiness, proneness to accidents (3%), eye or sight problems (3%), indecision, inefficiency (2%), cold sweats (2%)

An analysis of the data showed there to be three groups of women with a particular tendency to experience premenstrual symptoms:

(a) women who have, or who have had, children, as opposed to those who have not;
(b) women with irregular menstrual cycles, as opposed to those with regular cycles (the interpretation of 'irregular' and 'regular' was left to the respondents themselves);
(c) women with menstrual bleedings usually lasting 7 or more days, as opposed to women with a shorter period of menstruation.

The two groups of women, those with premenstrual syndrome and those without, did not differ significantly from each other as far as any of the other variables were concerned – such as age, education, socio-economic standing and marital status.

THE STATISTICAL EXERCISE

The statistical exercise was undertaken in order to see if the data obtained in France could in any way suggest why some women experience one premenstrual problem whilst others experience different ones.

For this exercise it was necessary to reduce the number of symptoms studied. This was done by grouping symptoms together, the clustering being done on 'clinical', rather arbitrary, grounds. Five groups of symptoms were thus arrived at:

D1 – skin problems (one symptom)
D2 – aches and pains (three)
D3 – symptoms related to water retention (three)
D4 – tenseness, general uneasiness, and fatigue (two)
D5 – 'nervous' symptoms (three).

It was then necessary to calculate a score for each woman in respect of each of these groups of symptoms. This was done by multiplying each complaint first by the number of days on which it was normally experienced, and then by the degree to which it was troubling, one for little, two for moderately, three for very much.

The next stage, and the main part of the exercise, was to fix the position of each woman in a five-dimensional space, each of the dimensions being governed by one of the groups of symptoms.

The result was the emergence of three very distinct and separate groups of women, outside of which very few were to be found.

Group A was governed primarily by D3 (symptoms related to water retention) and secondarily by D4 (general uneasiness and fatigue), all other symptoms being far less important (Table 3).

Group B was governed by D5 ('nervous' symptoms). The water retention symptoms (D3) and aches and pains (D2) were moderately important, but far less so.

Group C was governed by D2 (aches and pains), with D5 ('nervous' symptoms) playing a secondary and far less important role.

Table 3 Co-ordinates of the centres of gravity of the three groups

Group	*D1*	*D2*	*D3*	*D4*	*D5*	*No. women*	%
A	0.046	0.059	*0.572*	*0.330*	0.090	410	31.6
B	0.039	0.109	0.148	0.088	*0.624*	554	42.8
C	0.033	*0.771*	0.026	0.033	0.157	331	25.6

D1, skin problems, did not feature predominantly in any group. An analysis of the women who suffered skin problems showed that this was primarily a complaint of women over the age of 48; one wonders if this is not, in fact, a climacteric symptom rather than a premenstrual one. This is purely a matter of conjecture, however.

The next question was: Are there any explanations in the data available which might 'explain' why a woman is in one group as opposed to another, or rather, given the nature of the data available, are there any factors which dominate in any of the groups which might indicate the likelihood of a woman being in that group as opposed to any of the others? The only factors available were:

(a) socio-demographic factors: age, profession, husband's profession, parity, socio-economic standing, area of residence;
(b) physiological factors: length of menstrual cycle, duration of menstrual bleeding, cycle regularity, contraceptive use.

A segmentation analysis (Williams and Lambert, 1966) was used for this part of the exercise, and the analysis was based on 1064 respondents, all those who had provided information on all possible variables. A detailed description of the segmentation analysis procedure would be outside the scope of this paper. Suffice it to say that the computer programme selects the most discriminating variable, and indicates how the initial segmentation should be made. The process continues in each sub-group until no further statistical differences can be found or until the resulting sub-groups would consist of fewer than 20 respondents.

On this occasion the most important factor in determining whether a woman was in group A, B or C was found to be whether or not she had (or had had) children (Figure 1). Taking first the women without children (401, 38%), 27% belonged to group A, 39% to group B and 34% to group C. The figures for women with children (663, 62%) were 36%, 44% and 20% respectively. In other words, the figures being relative in the two groups, nulliparity increased a woman's chances to be in group C, whilst the having of children increased the likelihood of her being in groups A or B.

The second most important factor, both for nulliparous

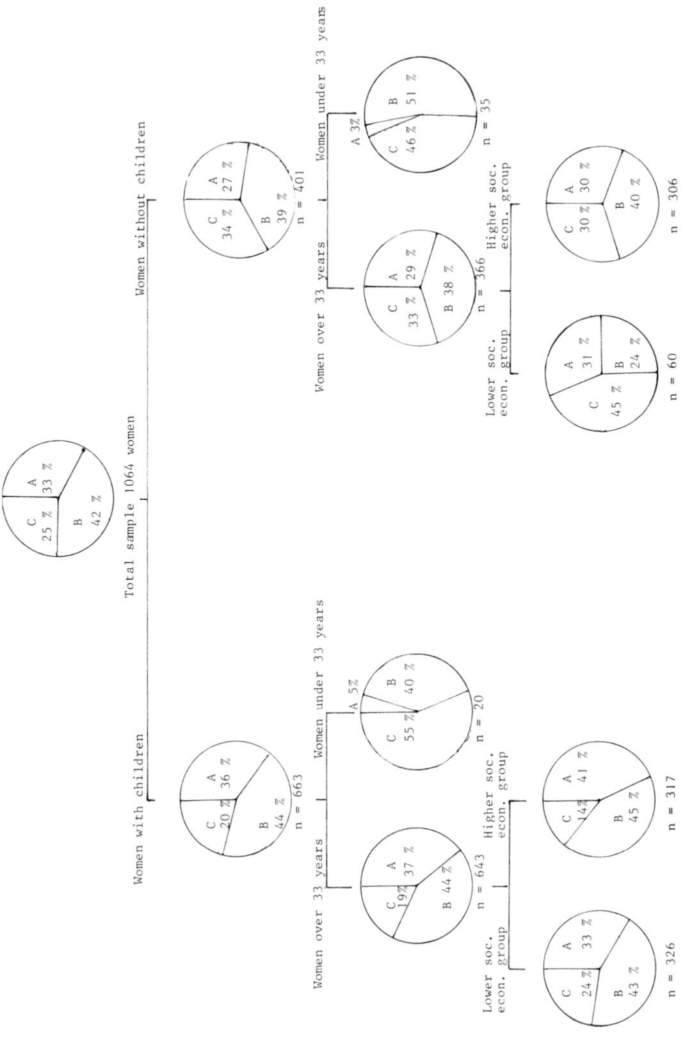

Figure 1 Results of segmentation analysis. A = symptoms related to water retention and secondarily to general uneasiness and fatigue; B = 'nervous' complaints; C = aches and pains

women and for those with children, was age. In both cases, women under 33 years of age were rarely found in group A, the group governed by D3, symptoms related to water retention, and D4, general uneasiness and fatigue. Thereafter, the profession, either of the woman herself or of her husband, became the influencing factor.

DISCUSSION

Among the facts to emerge from the French study were:

(a) 85% of the 2501 respondents had at one time or another experience of premenstrual syndrome;
(b) 77% still experienced premenstrual syndrome in one degree or another;
(c) 38% experienced premenstrual syndrome regularly every month.

These data are worth noting, especially in view of the fact that the study was conducted in a large unbiased sample, one which may be regarded as being representative of the French female population of fertile age.

The link between parity and an increased tendency to suffer premenstrual syndrome is not new, but the links between PMS and irregular menstrual cycles and prolonged menstrual bleeding seem not to have been reported before.

Turning to the statistical, heuristic, exercise carried out on the French data, the surprising emergence of three such distinct groups of women leads to the speculation that it is perhaps not *one* premenstrual syndrome with which we are dealing – but *three*. It would not be unthinkable that each of these three syndromes has its own aetiology, and, therefore, that each requires a different therapy. This might explain the variety in the ways in which women experience the premenstrual syndrome. It might also explain why some women react favourably to a given therapy, whilst others do not.

An alternative explanation could be that there is only one underlying disorder, but that environmental factors determine, in part, the way in which this is clinically expressed.

It is a great pity that more explaining variables were not available for analysis. It would indeed be very interesting to conduct a similar study, similar in size and character that is, and to include for analytical purposes the endocrine profiles of the respondents. It is likely that such a study would indeed provide good insight into the cause of the premenstrual syndrome or syndromes. Unfortunately, however, it seems likely that such a study would be prohibitively expensive.

References

Bickers, W. and Wood, M. (1951). Premenstrual tension; rational treatment. *Tex. Rep. Biol. Med.*, **9**, 406

Clare, A. W. (1977). Psychological profiles of women complaining of premenstrual symptoms. *Curr. Med. Res. Opin.*, **4** (Suppl. 4), 23

Dalton, K. (1969). *The Menstrual Cycle.* (New York: Pantheon)

Dalton, K. (1977). *The Premenstrual Syndrome and Progesterone Therapy.* (London: Heinemann, and Chicago: Year Book Medical Publishers)

Dalton, K. (1980). Cyclical criminal acts in premenstrual syndrome. *Lancet*, **2**, 1070

Frank, R. T. (1931). The hormonal causes of premenstrual tension. *Arch. Neurol. Psychiatr.*, **26**, 1053

Herschberg, A. D. (1966). Données statistiques sur l'étiologie et les traitements des mastopathies prémenstruelles. *C. R. Soc. Franç. Gynéc.*, nr. 5, May 1966

International Health Foundation (1979). *The Premenstrual Syndrome.* (Brussels/Geneva: International Health Foundation)

van Keep, P. A. and Haspels, A. A. (1979). Het premenstruele syndroom, een epidemiologisch onderzoek. *J. Drug. Res.*, **4**, 568

van Keep, P. A. and Lehert, P. (1980). Le syndrome pré-menstruel; une étude épidémiologique en France. *Contracept. Fertil. Stéril.*, **8**, 775

Kessel, N. and Coppen, A. (1963). The prevalence of common menstrual symptoms. *Lancet*, **2**, 61

Lehert, P. and van Keep, P. A. (1980). La nature du syndrome prémenstruel: une classification sur la population française et recherche de variables explicatives. (Available as a manuscript from International Health Foundation, 8 avenue Don Bosco, 1150 Brussels, Belgium)

Morton, J. H., Additon, H., Addison, R. G., Hunt, L. and Sullivan, J. J. (1953). A clinical study of premenstrual tension. *Am. J. Obstet. Gynecol.*, **65**, 1182

Sutherland, H. and Stewart, I. (1965). A critical analysis of the premenstrual syndrome. *Lancet*, **1**, 1180

Wetzel, R. D., Reich, T., McLure, J. N. and Wald, J. A. (1975). Premenstrual affective syndrome and affective disorder. *Br. J. Psychiat.*, **127**, 219

Williams, W. F. and Lambert, J. M. (1966). Multivariate methods in plant ecology. *J. Ecol.*, **54**

3
Premenstrual syndrome – a holistic approach

L. DENNERSTEIN, C. SPENCER-GARDNER and G. BURROWS

THE PREMENSTRUAL SYNDROME

Is there a specific premenstrual syndrome?

Changes in women's feelings, behaviour and social interactions in relation to the menstrual cycle have been observed for centuries. The term 'premenstrual tension' was introduced by Frank (1931) to describe a specific and severe syndrome of 'indescribable tension and irritability' present from 10 to 7 days preceding menstruation and relieved by the onset of menses. Israel (1938) reported that women with premenstrual tension suffered in the premenstrual fortnight 'a cyclic alteration of personality', unreasonable emotional outbursts, headache and 'nymphomania'. Later authors such as Dalton (1964) more broadly defined a premenstrual syndrome as a cluster of symptoms, both psychological and physical, which appear episodically in relation to the phases of the menstrual cycle. Dalton (1980) has emphasized that a symptom-free phase of a minimum of 7 days per cycle is needed to distinguish the syndrome from 'menstrual distress', a condition in which symptoms increase during the premenstrual week but are present throughout the cycle.

Various physical, psychological and behavioural changes have been suggested to comprise the premenstrual syndrome. The most prominent symptoms complained of by a group of women with a symptom-free phase in the menstrual cycle were irritability, swelling, depression, painful or tender breasts, tension and mood swings (Haskett *et al.*, 1980).

The incidence of the premenstrual syndrome varies depending on the definition of the syndrome used, sampling procedure and method. Many of the epidemiological studies have been based on retrospective questioning despite reports (McCance *et al.*, 1937; May, 1976) that there was no relationship between retrospective interview reporting of mood changes and that found by daily records.

Dalton (1980) reported that approximately half of all newly referred patients to a premenstrual syndrome clinic are found *not* to be suffering from premenstrual syndrome according to a definition requiring 7 symptom-free days. Haskett *et al.* (1980) rejected from their study 80% of volunteers who had claimed severe premenstrual symptoms. These results suggest that more women label their complaints as being due to a 'premenstrual syndrome' than do investigators. In addition, the use of different definitions of the syndrome or the failure to even specify the definitions used may hamper the comparison of research findings.

AETIOLOGY

Biological

A biological basis is suggested by the finding that female primates observed in the field have cyclic irritability corresponding to their fertility cycle (Janowsky *et al.*, 1966).

Mechanisms proposed include a relative deficiency of progesterone (Frank, 1931; Israel, 1938), excessive aldosterone action (Dalton, 1964), increased prolactin levels (Horrobin, 1973), decreased pyridoxine (Brush, 1979), and prostaglandin activity (Wood and Jakubowicz, 1980).

Conflicting results have been reported from studies which have compared hormone levels of premenstrual syndrome sufferers with those of controls. There is also little agreement from clinical trials based on these hypotheses. This may at least partly reflect the criteria of selection of patients for such studies.

Interestingly, most uncontrolled or single-blind studies report high rates of success no matter what the substance studied. Double-blind studies have found placebo to be very effective in the therapy of premenstrual complaints. This suggests that psychological factors probably play an important role in the aetiology of the premenstrual syndrome and therapeutic regimes.

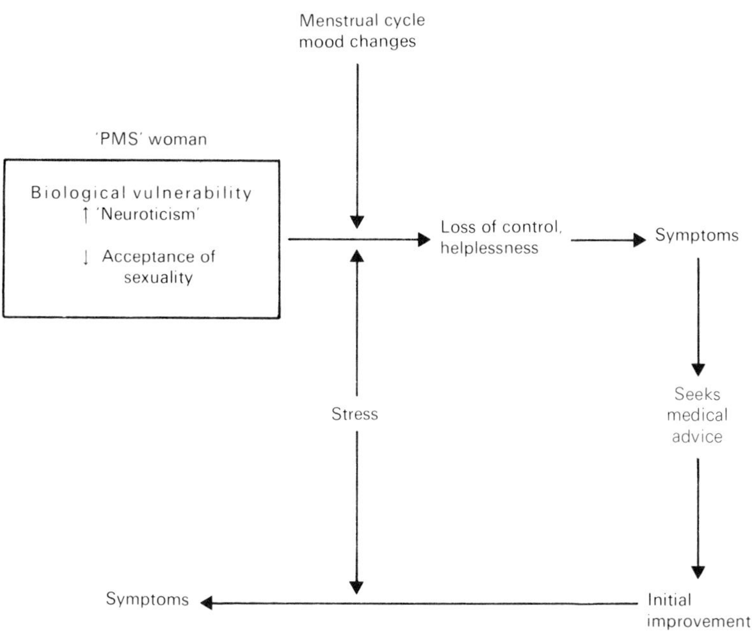

Figure 1 Factors in premenstrual tension

Psychological

There is an accumulating body of evidence relating to the possible aetiological roles of social learning (for example, of an association between menses and discomfort) and attribution by society of complaints to the menstrual cycle. The suggestion that premenstrual tension represents a rejection of femininity has not been confirmed in a recent study (Watts *et al.*, 1980), which did find that sufferers from the premenstrual syndrome had significantly raised trait anxiety and neuroticism scores and more negative attitudes to their bodies, genitals, sex and masturbation. Several studies link premenstrual tension with psychiatric ill-health. Women with emotional and sexual problems are more likely to report premenstrual complaints (Wood *et al.*, 1979). Psychiatric in-patients report significantly more severe distress on the Moos Menstrual Distress Questionnaire than do normal women (Zola *et al.*, 1979).

HYPOTHESES

It seems likely that there are some women who are unusually sensitive to endogenous hormonal changes and who present little evidence of intrapsychic or environmental stress as causative factors. Clinical experience suggests that these women form the minority of those who present with premenstrual complaints.

The biological profile of the average premenstrual syndrome sufferer suggests the following hypotheses: the woman most liable to suffer from premenstrual complaints is likely to be a vulnerable person by virtue of both biological and personality factors. High trait anxiety suggests a tendency to perceive situations as 'ego-threatening' (Shedletsky and Endler, 1974) with lowering of self-esteem or self-concept. High neuroticism scores are associated with irritability, nervousness, worry, mood changes, guilt and dissatisfaction (Tellegen, 1978). Biological vulnerability means she is likely to experience severe menstrual cycle mood fluctuations or other cyclical symptoms. This may reflect abnormalities of her hormonal milieu or of other substances, or even an individual sensitivity to normal hormone levels.

If these symptoms occur at the time of some other stress, feelings of loss of control and helplessness may be increased. As this woman has poor acceptance of her body image and sexuality, the genital organs may become the focus of fears. The woman then seeks medical advice, presenting as symptoms the menstrual-related changes which may have been exaggerated by her personality structure and inability to deal adequately with stress. Most therapies given are likely to be initially effective – especially if given with strong positive reassurance. The initial problems (i.e., 'stress' and personality factors) remain, and, if not dealt with adequately, symptomatology is likely to recur. The whole process may be exacerbated by attributional processes, such as increased society labelling of the premenstrual syndrome.

MANAGEMENT

Careful assessment is likely to prove the key to successful management. This involves delineation of the symptoms and thorough psychiatric assessment. Women suffering from major psychiatric syndromes should receive the appropriate therapy for that disorder and an appropriate explanation be given for the exacerbation of symptoms observed premenstrually. Attempts to attribute the psychiatric disorder to hormonal changes should be dissuaded. Daily rating of symptoms should then be carried out for 1 to 2 months and preferably be correlated with hormonal indicators of the menstrual cycle. This assessment will indicate the relationship of hormonal events to symptoms and, by encouraging introspection, may in itself be therapeutic.

As a result of assessment an appropriate therapy may be instituted. A group of women may be identified with problems present only in relation to specific hormonal phases of the menstrual cycle (usually premenstrual). The major type of problem varies between individuals. Breast swelling, nodularity and tenderness, headaches, mood changes and pain are among the specific problems which may be isolated to a cycle phase. Biological factors play a major role in the aetiology of these complaints, but psychological factors such as stress may make the symptoms intolerable and lead to the request for help. Therapy will depend on the type of complaint and the specific relationship with the menstrual cycle. Antiprostaglandins are effective for menstrual cramps and bromergocryptine for breast symptoms. When symptoms are oestrogen-related they may respond to a progestogen therapy throughout the month (days 5 to 25). Symptoms which correlate with the increase in progesterone in the luteal phase may respond to an oestrogen with a progestogen added monthly to induce shedding of the endometrium. Interestingly, some patients in this group have elected to have no pharmacotherapy after the assessment phase, claiming that they had learnt more about themselves and could now cope with the symptoms, for example, by reducing exposure to stress during the biologically vulnerable period or by increasing their coping repertoire.

Alternatively, the assessment may demonstrate symptoms at other phases of the menstrual cycle, with perhaps some premenstrual exacerbation. Interventions that may prove helpful for these women include those of decreasing intercurrent problems by environmental manipulation where possible, or offering treatment (of placebo and support) to those with little capacity for personality change with a strong desire for medication. Relaxation and cognitive therapy approaches, individual or group, may help these women learn techniques of coping more effectively with stress and increase their acceptance of sexuality so that the genitalia are less likely to be the focus of complaints.

The effectiveness of this management programme is currently being evaluated in an ongoing study of women with premenstrual complaints referred to the University of Melbourne, Department of Psychiatry.

References

Brush, M. G. (1979). Endocrine and other biochemical factors in the aetiology of the premenstrual syndrome. *Curr. Med. Res. Opin.*, **6** (Suppl. 5), 19

Dalton, K. (1964). *The Premenstrual Syndrome.* (London: Heinemann)

Dalton, K. (1980). Progesterone, fluid and electrolytes in premenstrual syndrome. *Br. Med. J.*, **281**, 61

Frank, R. H. T. (1931). The hormonal causes of premenstrual tension. *Arch. Neurol. Psychiatry*, **26**, 1053

Haskett, R. F., Steiner, M., Osmun, J. N. and Caroll, B. J. (1980). Severe premenstrual tension: delineation of the syndrome. *Biol. Psychiatry*, **15**, 121

Horrobin, D. F. (1973). *Prolactin: Physiology and Clinical Significance.* (Lancaster: MTP)

Israel, S. L. (1938). Premenstrual tension. *J. Am. Med. Assoc.*, **110**, 1721

Janowsky, D. S., Gorney, R. and Kelley, B. (1966). The curse – vicissitudes and variations of the female fertility cycle. *Psychosomatics*, **7**, 242

May, R. R. (1976). Mood shifts and the menstrual cycle. *J. Psychosom. Res.*, **20**, 125

McCance, R. A., Luff, M. C. and Widdowson, E. E. (1937). Physical and emotional periodicity in women. *J. Hyg. (London)*, **37**, 571

Shedletsky, R. and Endler, N. S. (1974). Anxiety: the state-trait model and the interaction model. *J. Pers.*, **42**, 511

Tellegen, A. (1978). Review of the EPI. In Buros, O. K. (ed.) *The Eighth Mental Measurements Year Book.* (New Jersey: Gryphon)

Watts, S., Dennerstein, L. and Horne, D. J. de L. (1980). The premenstrual syndrome: a psychological evaluation. *J. Affect. Disord.*, **2**, 257

Wood, C. and Jakubowicz, D. (1980). The treatment of premenstrual tension with mefenamic acid. *Br. J. Obstet. Gynaecol.*, **87**

Wood, C., Larsen, L. and Williams, R. (1979). Social and psychological factors in relation to premenstrual tension and menstrual pain. *Aust. N.Z. J. Obstet. Gynaecol.*, **19**, 111

Zola, P., Meyerson, A. T., Reznikoff, M., Thornton, J. C. and Concool, B. M. (1979). Menstrual symptomatology and psychiatric admission. *J. Psychosom. Res.*, **23**, 241

4
An appraisal of the role of progesterone in the therapy of premenstrual syndrome

G. A. SAMPSON

Progesterone is a steroid produced initially by the granulosa cells of the ovary. Small amounts are probably secreted into the centre of the follicle before the start of the LH surge, but it is unlikely that much of it gets into the general circulation due to the relatively impermeable basement membrane. Following the LH surge the follicle becomes vascularized and large amounts of progesterone and oestradiol are secreted into the ovarian vein. The combined action of oestrogen and of progesterone converts the late proliferative endometrium to the secretory one necessary for implantation of the blastocyst (Lenton and Cooke, 1974). Plasma progesterone levels rise from ovulation to the fifth postovulatory day and then plateau for 3 or 4 days before falling if conception has not occurred. Usual mid-luteal levels are considered to be above 5 ng/ml in an ovulatory cycle and above 9 ng/ml in a potentially fertile cycle (Lenton, E. A., unpublished data).

Progesterone cannot be administered orally as it passes by the portal system to the liver where it is rapidly metabolized; it has a half-life of 15 minutes. It has, therefore, to be given by rectal, vaginal or parenteral administration or by implantation into the fat of the abdominal wall. Nillius and Johansson (1971) showed that an intramuscular injection of 25 mg progesterone resulted in normal luteal phase plasma progesterone levels within 8 hours, and that 100 mg injections gave mid-pregnancy levels. These elevated levels were maintained for at least 48 hours. Progesterone 100 mg, given rectally or vaginally, produced luteal phase plasma progesterone levels, peak levels being reached within 4 hours after vaginal administration and 8 hours after rectal administration; there was then a gradual decline, reaching follicular phase levels after 24 hours. They suggest that during absorption much of the progesterone diffuses into the fatty tissue and diffuses back into the blood stream when the progesterone levels begin to fall.

Premenstrual syndrome is a global term which indicates changes in mood, behaviour and physical symptoms in relation to the menstrual cycle. Classically, there is an increase in symptoms which reach a peak in the late luteal phase and then diminish in intensity with the onset of menstruation. There are many definitions (Dalton, 1977; Rees, 1953) listing different symptoms, but these are primarily irritability, depression, breast tenderness, bloating, difficulty in concentrating and behavioural changes. The essence of any definition of premenstrual syndrome is the periodicity of the symptoms and their timing in relation to the onset of menstruation. Incidence studies suggest that over 50% of menstruating women show some cyclical changes in mood, feeling swollen, or behaviour in relation to menstruation; it is perhaps abnormal not to show such changes. We therefore have to differentiate between those who suffer from premenstrual syndrome and those who have 'normal' changes premenstrually. The question of whether premenstrual syndrome is just one extreme of a normal tendency or whether it is a clear-cut disease entity has not yet been answered. My personal definition of a 'patient' with premenstrual syndrome includes facts such as that her symptoms are marked and affect her own or other people's lives, and that she is actively seeking help.

Katharina Dalton first suggested a relationship between progesterone and premenstrual syndrome and has since written many papers and a monograph on the subject. Her initial suggestion was based on clinical observation, but since it has been possible to assay progesterone levels several workers have attempted to find a direct relationship between progesterone and premenstrual syndrome.

Bäckström and Carstensen (1974) found that in women who have mainly psychological symptoms oestrogen levels are raised and progesterone levels reduced in plasma during the week preceding menstruation as compared to a 'normal' group of women.

O'Brien *et al.* (1980) reported that plasma progesterone concentration is higher in women with symptoms during the post-ovulatory phase of the cycle and the peak progesterone concentration appears earlier.

Munday (1977) found a deficit of progesterone 8 days premenstrually in eight women with severe premenstrual syndrome; interestingly, in six of the eight women studied symptoms started more than 8 days premenstrually. In a larger study of 58 women she found lower progesterone levels in the second half of the cycle in only 30% of subjects.

Smith (1975) found the mean plasma levels of progesterone in the 7 days preceding menstruation lower in 18 patients over 24 cycles than in 11 controls over 14 cycles, the difference being statistically significant although small. Smith was impressed by the fact that there is a tremendous overlap between patient and control group levels, and also that for any given individual the presence and degree of depression in any given cycle is unrelated to the particular levels of hormone during that cycle.

In Sheffield we studied untreated and progesterone-treated cycles in five patients in a metabolic unit. Mood was assessed by the Moos Menstrual Distress Questionnaire (Moos, 1977), and weight, basal body temperature and measures of swelling and oedema were recorded daily. Daily 24 h urine samples were collected and urinary electrolytes, androsterone, etiocholanolone, dehydroepiandrosterone, 11 oxoandrosterone, 11 oxoetiocholanolone, pregnanolone, 11β-hydroxyandrosterone, 11β-hydroxyetiocholanolone, allo-pregnanediol and pregnanediol were measured. Plasma samples were assayed for progesterone and 17β-oestradiol. We found no consistent pattern of hormonal or electrolyte change when comparing untreated cycles. In two untreated cycles progesterone levels were low, reaching a peak of 8 ng/ml; in two, levels were in the upper part of the normal range and in one the luteal phase plasma progesterone rise was absent, although an apparent luteal 17β-oestradiol rise was observed (Fenoughty, 1975).

It seems at present that we have no good evidence that patients suffering from premenstrual syndrome have abnormal progesterone levels, and hormone studies offer no rational basis for progesterone therapy unless an individual patient can be shown to have progesterone deficiency.

The initial observation that there is a relationship between premenstrual syndrome and progesterone was, however, a clinical one. Dalton has used progesterone for nearly 30 years in the treatment of premenstrual syndrome. She feels failure of patients with premenstrual syndrome to respond is due either to (a) incorrect diagnosis or (b) incorrect frequency or dosage of progesterone.

Dalton suggests a dosage range from 100 mg daily to 400 mg t.d.s. by suppository or pessary with a mean of 400 mg daily; and from 50 mg on alternate days to 100 mg daily by intramuscular injection. She advises that it is wise to start a course of progesterone treatment 5 days before symptoms are expected; most patients require treatment from day 14, but there is a range from day 8 to day 28. If patients become symptom-free the dosage and duration of treatment may be reduced. Progesterone pellets can be implanted into the fat of the anterior abdominal wall, the usual dose being 5 to 12 × 100 mg pellets. In a series of 100 consecutive hospital patients suffering from severe premenstrual syndrome she reports that suppositories or pessaries (mean dosage 400 mg/day from day 14) were beneficial to 75 patients, injections to nine, and an implant to two patients. Fourteen patients became symptom-free with treatment and were able to stop treatment.

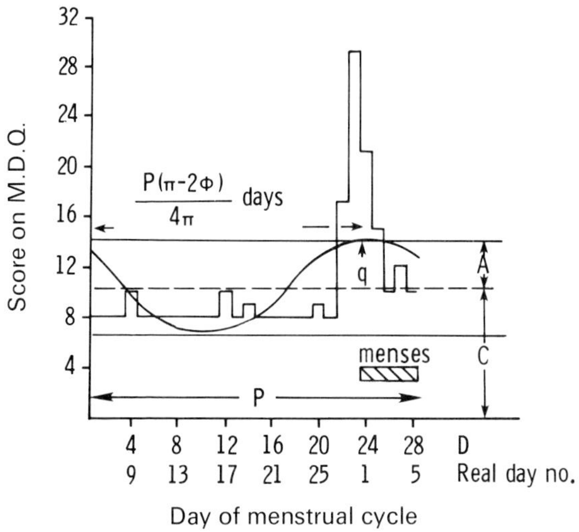

Figure 1 Data showing the negative affect scores on the Moos Menstrual Distress Questionnaire (MDQ) of one patient throughout a menstrual cycle. The least mean square fitted sine wave is also shown, and the significance of the various constants is illustrated. The sine wave is the one for which the sum of the squares of the deviations of the sine wave from the crude data is minimal

Rees (1953) reported that progesterone therapy was beneficial; Gray (1941) claimed 35 of 38 patients were greatly improved, but progesterone was given in low dosage. Smith (1975) quoted 'in our own clinic many patients have been found who seem to have derived benefits from this treatment'. In an uncontrolled study of progesterone treatment in 40 patients in the Sheffield clinic, 20 (50%) reported it helpful, 14 (35%) said it made their symptoms worse and 6 (15%) felt it produced no change.

Uncontrolled studies or reports from clinics suggest that progesterone is helpful in premenstrual syndrome. However, when reviewing the treatment of premenstrual syndrome by any therapy given in an uncontrolled study one finds impressive results. When the treatment is given in a controlled double-blind study there is usually (a) a high placebo response and (b) the active agent is usually no better than placebo. The presence of a high placebo response in a disorder of affect is not surprising and the problems in assessing response to therapy parallel those in assessing antidepressant and anxiolytic drugs.

Two double-blind studies comparing placebo and progesterone have been reported. Smith (1975) reported a study of 14 patients treated over four cycles comparing the effectiveness of four regimes: progesterone, progesterone and spironolactone, spironolactone, and placebo. Progesterone was given intramuscularly, 50 mg on alternate days from day 19. The results indicate that there were three patients who did better in progesterone months, three who did better in non-progesterone months, and eight whose results appeared to bear no relationship to the presence or absence of progesterone. They measured plasma levels of progesterone and oestrogen and found levels of hormones to be unrelated to therapeutic benefit or lack of it.

Sampson (1978) completed a double-blind controlled trial of progesterone given by suppository or pessary compared with placebo. Thirty-nine patients with a mean age of 34.7 years, 82% of whom were parous, were studied. The first phase of the trial consisted of a recorded untreated cycle, and then a 2-month crossover of placebo and progesterone 200 mg b.d. given from day 16 to 26 of a standard 28-day cycle but with the timing of dosage adjusted for cycle length. The second phase of the trial, completed by 26 patients who had already completed the first phase, was a double-blind crossover of placebo and progesterone 400 mg b.d. Symptoms were rated by both daily menstrual distress questionnaires and by retrospective reporting. On reviewing the patients' retrospective reports recorded after every cycle, when progesterone 200 mg b.d. was compared to placebo, 31% reported that progesterone was better than placebo and 43% that placebo was better than progesterone; 20% found neither helpful. When progesterone 400 mg b.d. was compared to placebo, 27% found progesterone more helpful than placebo, 35% found placebo more helpful than progesterone, and 23% reported that neither was helpful.

The Moos Menstrual Distress Questionnaire was completed daily, and the daily scores transformed into symptom scales assessing pain, concentration, behavioural change, autonomic response, water retention, negative affect, arousal and control. In order to assess a significant rise in symptoms in the premenstruum the Best Fit Sine Wave Method (Sampson and Jenner, 1977) was applied to all symptom scales (Figure 1). This gives an 'A' value for each symptom scale in each cycle which is an estimate of the amplitude of complaining in relation to menstruation. The 'q' value (acrophase) gives an indication of the timing of the symptoms in relation to the onset of menstruation and only 'A' values where the acrophase was between 210 and 330 degrees, i.e. day 23 to plus three of a 28-day cycle, were considered to be indicative of premenstrual symptoms. The presence of an 'A' indicates a significant premenstrual increase in symptoms, therefore reduction in the number of significant 'A' values in treated cycles indicates effectiveness of therapy.

Figure 2 shows the number of cycles without significant 'A' values. Non-parametric statistics (chi square) give an assessment of the difference between cycles, as shown in Table 1. Interestingly, neither progesterone- nor placebo-treated cycles were significantly different from untreated cycles in the autonomic reaction and water retention scales. Symptoms in the negative affect, behavioural change and control scales were significantly reduced by progesterone 200 mg b.d. and placebo, but there was no significant difference between the response to placebo or progesterone. Symptoms in the concentration scale were significantly improved in the first placebo cycle and by progesterone 200 mg b.d. The symptoms in the pain scale, which include headache, backache, cramps and fatigue, were significantly diminished by placebo and progesterone 200 mg b.d., but when progesterone 400 mg b.d. is compared with the second placebo cycle the placebo cycle was significantly more effective ($p<0.01$). The results show that both progesterone and placebo are significantly effective in reducing symptoms,

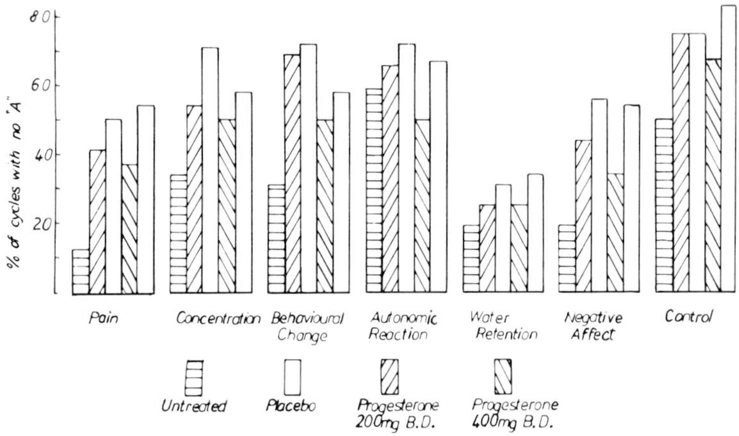

Figure 2 The percentages of untreated, progesterone-treated and placebo-treated cycles in which there is no 'A' (a measure of complaining in relation to menstruation) for each scale of the Moos Menstrual Distress Questionnaire

especially in the pain, concentration, behavioural change and negative affect scales of the Moos Menstrual Distress Questionnaire. Placebo is usually more effective, but there is not a significant difference between placebo and progesterone therapy.

The most common problem in the study was the dislike of vaginal or rectal insertion, but no patient dropped out because of this. Premenstrual spotting occurred in one treated cycle in a patient with an IUD. Four patients developed monilia, but all had had infections previously.

Table 1 Significant differences between untreated, placebo-treated and progesterone-treated cycles when comparing numbers of 'A' (a measure of complaining in relation to menstruation)

	Untreated	*Untreated*	*Untreated*	*Untreated*	*Progesterone 200 mg b.d.*	*Progesterone 400 mg b.d.*
	Progesterone 200 mg b.d.	*Progesterone 400 mg b.d.*	*Placebo (i)*	*Placebo (ii)*	*Placebo (i)*	*Placebo (ii)*
Pain	●		●●	●		●●
Concentration	●		●●			
Behavioural change	●●		●●	●●		
Autonomic reaction						
Water retention						
Negative affect	●		●●	●●		
Control	●		●	●●		

● = $p < 0.05$ ●● = $p < 0.01$

The apparent effectiveness of the placebo over progesterone 400 mg b.d. in the pain scale suggests that the progesterone in a larger dose may be increasing some of the symptoms. This apparent increase in symptoms occurs premenstrually and not with menstruation. The pain scale of the Moos Menstrual Distress Questionnaire records the following symptoms: muscle stiffness, headache, cramps (uterine or pelvic), backache, fatigue and general aches and pains. If we calculate the mean 'q' value in days for cycles with a significant 'A' for pain, i.e. the timing of the symptom peak, we find that the number of days before menstruation in the untreated cycle is 3.14 days, in the progesterone 200 mg b.d. treated cycle 1.79 days, in the first placebo cycle 2.38 days, in the second placebo cycle 1.45 days and in the progesterone 400 mg b.d. cycle 5.1 days. All these times are clearly premenstrual; the higher dose of progesterone produces symptoms earlier in the cycle and the effect does not seem to be solely explained by increased dysmenorrhoea. Pain symptoms in the untreated cycle peak 3 days before menstruation, and this confirms that our patient group had premenstrual syndrome rather than spasmodic dysmenorrhoea.

In summary, both double-blind trials found that progesterone can be effective in reducing premenstrual symptoms, but that placebo is equally effective.

These findings are in accord with Somerville's study on menstrual migraine (1972), where he found that treatment with progesterone had no effect on the migraine of four out of six patients, that it reduced the severity and duration in one patient, and that it prevented the attack in the remaining patient. Withdrawal of progesterone did not initiate migraine.

In controlled clinical trials it appears that administration of progesterone in the luteal phase is no more successful than administration of placebo. There are, however, many patients who claim that progesterone therapy has 'revolutionized' their lives. In an attempt to assess this claim I have reviewed patients in the Sheffield clinic who are of this opinion.

Patient L, aged 37, developed recurrent episodes of depression, anergia, loss of appetite, difficulty in concentrating, sleep disturbance, headaches, anxiety and bloated feeling at the age of 34 years. In the 6 months prior to the onset of symptoms she had a haemorrhoidectomy, a total dental extraction and an illness diagnosed as glandular fever. Her symptoms were so marked that she had been admitted to a psychiatric hospital on several occasions and received antidepressants, anxiolytics and ECT, with no benefit. Her menarche was at 14 years; she had regular pain-free menses and no mood changes with menstruation. She had two pregnancies, the first at 25 years resulted in a stillbirth, the second at 30 years was a full-term normal delivery. She had no pre-eclamptic toxaemia, postnatal depression or 3-day 'blues' with either pregnancy. Menstruation returned regularly after the second pregnancy and for 4 years was uneventful. There was no family history of psychiatric illness or premenstrual syndrome; she had a stable marriage, and we could find no evidence of psychological stress sufficient to induce her symptoms. The patient was eventually admitted to a metabolic unit and studied over an untreated and a progesterone-treated cycle, progesterone being given 100 mg i.m. daily from day 16 to 28. Figure 3 shows daily scores on the eight scales of the Moos Menstrual Distress Questionnaire and her plasma progesterone levels. Her behavioural change and negative affect scales were significantly improved in the treated cycle. Her fluid retention symptoms began on day 10 of the treated cycle, although they had not started until day 14 of the untreated cycle. In the second cycle there was a rise in plasma progesterone by day 8 of the cycle; there was, however, no corresponding ovulatory 17β-oestradiol peak. During progesterone therapy normal luteal phase plasma progesterone levels were maintained, but during the same time the oestrogen : progesterone ratio was consistently low. Although the plasma progesterone levels in the treated cycle appeared to reach twice the value of those in the untreated cycle, urinary pregnanediol levels remained similar in both cycles, as did levels of allo-pregnanediol. Urinary pregnanolone was excreted at levels approximately twice as high during the treated cycle as during the untreated cycle.

This patient's clinical progress has been recorded over an 8-year period; the first 3 years are illustrated in Figure 4. She felt that progesterone 100 mg i.m. from day 16 to day 28 of her cycle

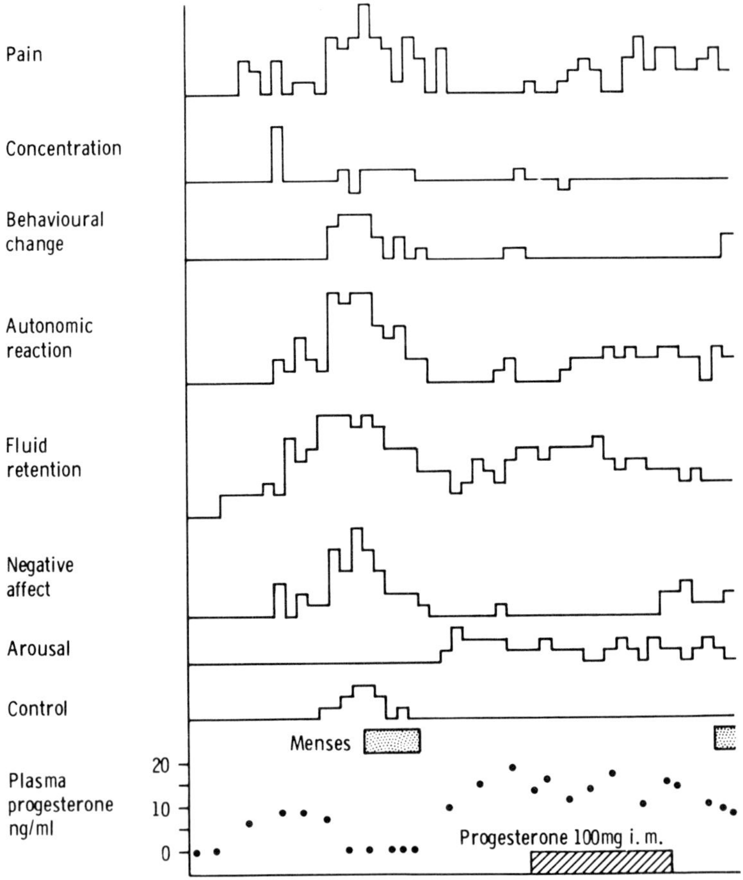

Figure 3 Daily Moos Menstrual Distress symptom scores and plasma progesterone levels in an untreated and in a progesterone-treated cycle

reduced the intensity and frequency of her symptoms, but it also altered her cycle pattern from a regular 24/26-day cycle to an irregular cycle with a longer length. She soon developed problems with heavy continuous bleeding and after 5 calendar months her therapy was altered to progesterone p.v., 1000 mg b.d., progesterone 50 mg i.m. on alternate days from day 17 to 27 of a superimposed 28-day cycle. This virtually suppressed menstruation for 8 calendar months and also reduced the number of days with symptoms. She then, with no change in therapy, began to bleed with a 26/28-day cycle and symptoms

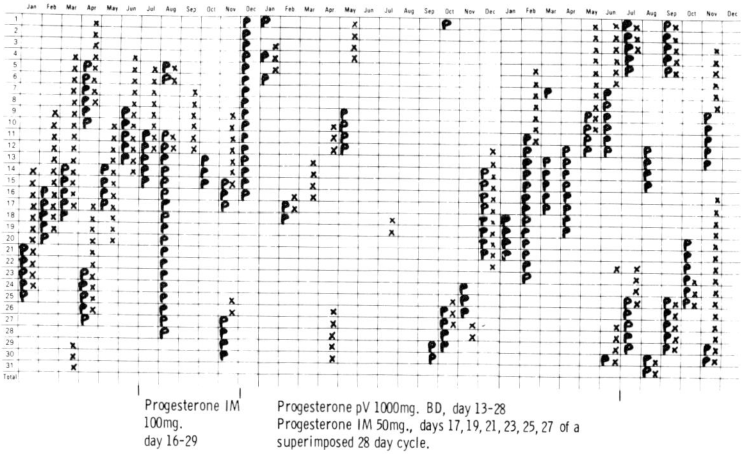

Figure 4 Chart showing menstruation (P) and symptoms (X) of one patient (L) over 3 years

recurred. However, bleeding could occur without symptoms, but after 3 months of her initial symptoms with every bleed all medication was stopped. Her cycle reverted to a 24/26-day pattern; she continued to have symptoms. After 4 untreated calendar months we gave her the contraceptive pill continuously in an attempt to suppress menstruation, and she took this pill (ethinyloestradiol 30 μg, levonorgestrel 250 μg) for 14 days.

After 3 days of treatment she became depressed and this depression lifted with the onset of bleeding 4 days after stopping the pill. After this time she had 68 days with no bleeding and no symptoms. Ultimately we suppressed the patient's menstrual cycle with 17 hydroxy-progesterone hexanoate and this has stopped both bleeding and symptoms.

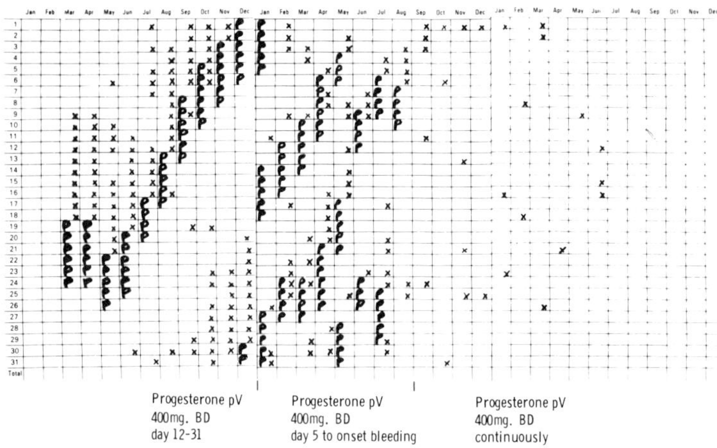

Figure 5 Chart showing menstruation (P) and symptoms (X) of one patient (M) over 3 years

Progesterone therefore helped this patient's symptoms, but it also disrupted her normal cycle pattern. The more effectively her cycle was disrupted the more effectively were her symptoms controlled. She also demonstrates that premenstrual symptoms can be dissociated from bleeding. She required a high dosage of progesterone to control symptoms and alter her cycle pattern.

Patient M is 39 years old and married. Her menarche was at 13 years and she was poorly prepared for it. Her early menstruation was irregular with mild dysmenorrhoea. Her premenstrual symptoms became marked in her mid-thirties with 10 days of depression, irritability and insomnia, but her symptoms stopped with the onset of bleeding. She has had three pregnancies, two full-term normal deliveries with no postnatal depression, and a termination followed by sterilization at the age of 34 years. At first she received the standard regime of progesterone p.v. 400 mg b.d. from day 12 to day 31 of her cycle.

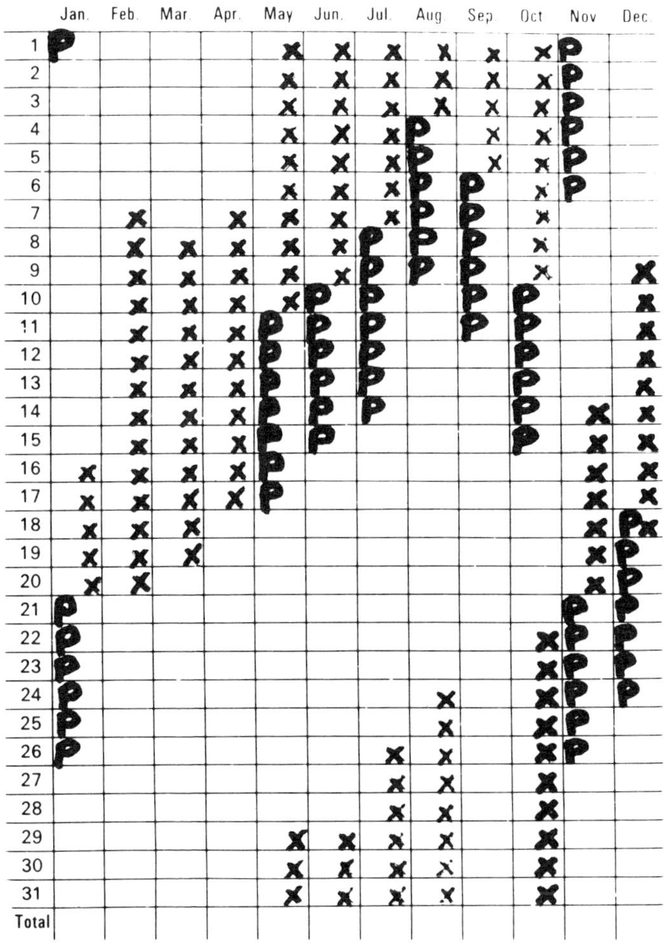

Figure 6 Chart showing menstruation (P) and symptoms (X) of one patient (T) when progesterone 400 mg noct. p.v. was given daily for 82 days from 25th January to 17th April

This did not reduce the frequency of symptoms; she reported that it reduced the intensity of her symptoms but her family did not agree that it did. An attempt to alter this patient's cycle pattern by affecting ovulation was made by giving progesterone in the follicular phase; she therefore received progesterone p.v. 400 mg b.d. from day 5 to the onset of bleeding. This regime altered her cycle length and the number of days with symptoms (Figure 5). After 4 months of this regime she failed to bleed and remained continuously on progesterone 400 mg b.d. As Table 2 shows, progesterone appears to be effective only when the normal cycle pattern is disrupted.

Table 2 The number of days of symptoms (X) and of bleeding (P) per calendar month in one patient during various indicated treatment regimes

Treatment	*Number of cycles*	*Cycle length*	*'X' per calendar month*	*'P' per calendar month*
Untreated	4	31	10	6
Progesterone p.v. 400 mg b.d. day 12–31	6	27	11	7
Progesterone p.v. 400 mg b.d. day 5 to onset of bleeding	8	14	7	14
Progesterone p.v. 400 mg b.d. continuously	—	—	3	0

In patient T we tried an identical regime; this stopped bleeding, but the premenstrual symptoms continued in their regular pattern (Figure 6), again illustrating that symptoms can occur without bleeding in some patients.

DISCUSSION

The effectiveness of progesterone in treating premenstrual symptoms should be compared with other therapies. In uncontrolled trials the effectiveness of therapy is often amazing, ranging from 91% for ammonium nitrate 1 g t.d.s. (Steiglitz and Kimble, 1949) and 90% for vitamin A 200000 units per day (Block, 1960) to a more usual 50%. Most studies using placebo find that it has at least a 50% success rate, a value which would be expected in studies treating a psychosomatic disorder.

Rees (1976) described a working model of a possible mechanism involved in the premenstrual syndrome: 'The syndrome is primarily a somatopsychic syndrome in which the physical changes are primary, but the degree of distress and disability experienced will be affected by the emotional state of the patient. If the patient suffers from a neurotic illness or has traits of personality indicative of over anxious, obsessional or hypochondriacal tendencies the degree of distress and disability associated with the premenstrual syndrome can be increased. The severity of the syndrome will be the product of the degree of changes in the internal environment and the reaction of the patient as determined by her personality, stability and emotional state.'

Any rationale for progesterone therapy for premenstrual syndrome must be to alter what Rees describes as the 'internal environment'. Although progesterone obviously can have an effect on the 'reaction of the patient', it is difficult to ascribe this to anything except a placebo effect, i.e. progesterone, although it has some sedative properties, is not an anxiolytic, analgesic or antidepressant.

The double-blind controlled trials suggest that progesterone given in the luteal phase is as effective as placebo, and this would suggest it is affecting the 'reaction of the patient' and not the 'internal environment'. Placebo is as effective and a cheaper and perhaps safer way of affecting the 'reaction of the patient'.

In the individual case studies described it appears that progesterone when it does not alter the menstrual cycle pattern is not effective. However, when the menstrual cycle pattern is disrupted there is usually, but not always, an improvement in symptoms. In such an instance I would feel that progesterone is altering 'the degree of changes in the internal environment', as described by Rees. In the same paper, Rees described a case where therapy with hydroxy-progesterone hexanoate or oral progestogens was helpful, but the most benefit was obtained with a non-virilizing androgen. Dalton has described how progesterone implants, which she finds effective, inhibit ovulation for many months and produce irregular menstruation, scanty loss, or spells of amenorrhoea lasting 6 months, i.e. progesterone is effective because it disrupts the normal cycle pattern and alters the 'internal environment'. Progesterone obviously can alter the 'internal environment'; however, one should assess if other methods of disrupting the normal cycle are not safer, cheaper and/or more reliable in their effectiveness.

We do not yet know what, if any, abnormal endocrinology is the abnormal 'internal environment' which Rees and other workers have suggested as the primary change in premenstrual syndrome upon which many secondary factors impinge. It may not necessarily be abnormal progesterone levels, and a direct elevating of progesterone levels does not alleviate symptoms. Progesterone appears to be successful when the usual menstrual cycle pattern is disrupted.

References

Bäckström, T. and Carstensen, H. (1974). Estrogen and progesterone in plasma in relation to premenstrual tension. *J. Steroid Biochem.*, **5**, 257

Block, E. (1960). The use of vitamin A in premenstrual tension. *Acta Obstet. Gynaecol. Scand.*, **39**, 586

Dalton, K. (1977). *The Premenstrual Syndrome and Progesterone Therapy*. (London: Heinemann)

Fenoughty, M. (1975). The development and application of steroid assay procedures in a study of premenstrual syndrome. *Master of Medical Science Thesis*, University of Sheffield, UK

Gray, L. A. (1941). The use of progesterone in nervous tension states. *Southern Med. J.*, **34**, 1004

Lenton, E. and Cooke, I. D. (1974). Other disorders of ovulation. *Clin. Obstet. Gynaecol.*, **1**, 313

Moos, R. H. (1977). *Menstrual Distress Questionnaire Manual.* Social Ecology Laboratory, Department of Psychiatry and Behavioral Sciences, Stamford University, California, USA

Munday, M. (1977). Hormone levels in severe premenstrual tension. *Curr. Med. Res. Opin.*, **4** (Suppl. 9), 16

Nillius, S. J. and Johansson, E. D. B. (1971). Plasma levels of progesterone after vaginal, rectal or intramuscular administration of progesterone. *Am. J. Obstet. Gynecol.*, **110**, 470

O'Brien, P. M. S., Selby,, C. and Symonds, E. M. (1980). Progesterone, fluid and electrolytes in premenstrual syndrome. *Br. Med. J.*, **280**, 1161

Rees, L. (1953). Psychosomatic aspects of the premenstrual tension syndrome. *J. Ment. Sci.*, **99**, 62

Rees, W. L. (1976). Stress, distress and disease. *Br. J. Psychiatr.*, **128**, 3

Sampson, G. A. (1979). Premenstrual syndrome: a double-blind controlled trial of progesterone and placebo. *Br. J. Psychiatr.*, **135**, 209

Sampson, G. A. and Jenner, F. A. (1977). Studies of daily recordings from the Moos Menstrual Distress Questionnaire. *Br. J. Psychiatr.*, **130**, 265

Smith, S. L. (1975). Mood and the menstrual cycle. In Sachar, E. J. (ed.) *Topics in Endocrinology.* (New York: Grune and Stratton)

Somerville, B. W. (1972). The influence of progesterone and estradiol upon migraine. *Headache*, **12**, 93

Stieglitz, E. J. and Kimble, S. T. (1949). Premenstrual intoxication. *Am. J. Med. Sci.*, **218**, 616

5
An explorative study into the clinical effects of dydrogesterone in the treatment of premenstrual syndrome

J. R. STRECKER

INTRODUCTION

From the endocrinological standpoint it seems likely that a deficiency of progesterone in relation to 17β-oestradiol during the second half of the menstrual cycle plays an important role in the aetiology of the premenstrual syndrome.

Progesterone is synthesized exclusively in the corpus luteum during the second half of the cycle, and many patients suffering premenstrual syndrome are found to have an impaired corpus luteum function. This impairment may be caused by an elevation, even a moderate one, of serum prolactin and/or prostaglandin. The regulatory effects of these hormones on corpus luteum function has been shown by several studies (Litschgi and Glatthaar, 1978; Benedek-Jaszmann and Hearn-Sturtevant, 1976; Timonen and Procopé, 1973).

If this is so, it is reasonable to theorize that progesterone supplementation by pharmacological means during the second half of the cycle may well alleviate, if not completely resolve, premenstrual complaints. Accordingly, dydrogesterone, a stereo-isomer of progesterone, was administered to a group of patients with diverse but severe premenstrual complaints. The results of this clinical investigation are presented here.

MATERIAL AND METHODS

Thirty-one patients were treated with dydrogesterone (Duphaston®). This stereo-isomer of progesterone has similar effects to those of progesterone, but as it does not inhibit the gonadotrophins it does not prevent ovulation. The dosage was 20 mg/day during the second half of the menstrual cycle, from day 15 to day 25.

The patients were suffering various of the broad range of symptoms usually regarded as being part of the premenstrual syndrome. In the figures which follow only the most severe problems of each patient are reported upon; in most cases, one complaint per patient. Some women were also suffering menstrual problems, such as dysmenorrhoea and hypermenorrhoea. The ages of the patients ranged from 19 to 49.

The women were treated in the Out-patients Department of the Frauenklinik of the Universität Ulm, West Germany. They were treated for a minimum of three cycles, during which time they were seen at least four times.

RESULTS

Four premenstrual symptoms were considered in this study: depression, which 35% of the women taking part in this investigation reported as being a serious problem, oedema (29%), headaches (27%) and breast tenderness (23%).

Eleven patients were suffering badly from premenstrual depression. After 3 months of treatment with 20 mg dydrogesterone per day throughout the second half of the menstrual cycle, an improvement was found in six of these patients. In four the problem remained unchanged, and in one it deteriorated (Figure 1).

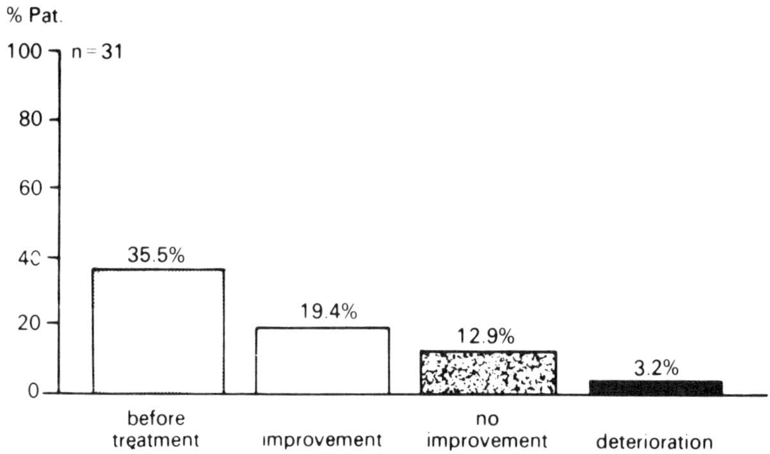

Figure 1 Premenstrual depression and the effect of therapy

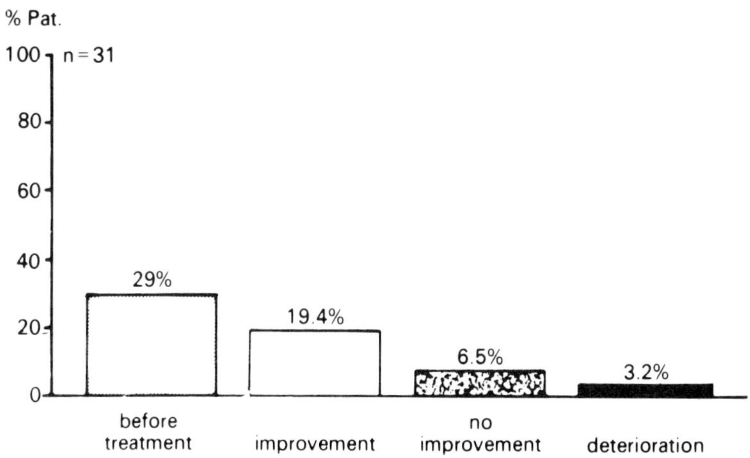

Figure 2 Premenstrual oedema and the effect of therapy

Nine patients (29%) complained of premenstrual oedema. After therapy, six showed an improvement. Two showed no change, and one got worse (Figure 2).

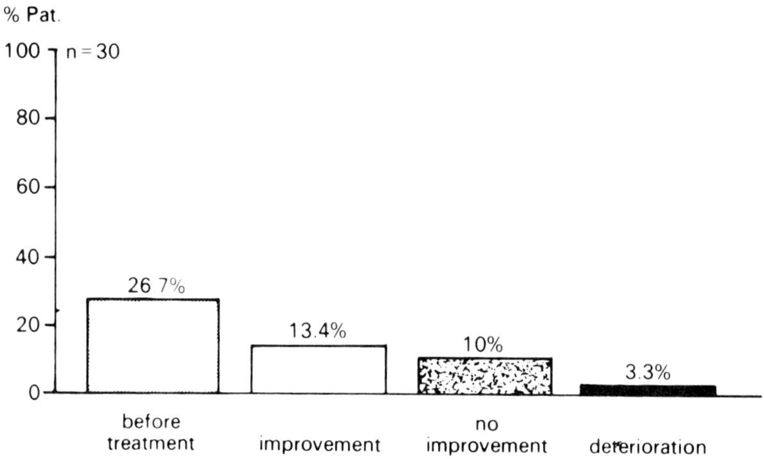

Figure 3 Premenstrual headache and the effect of therapy

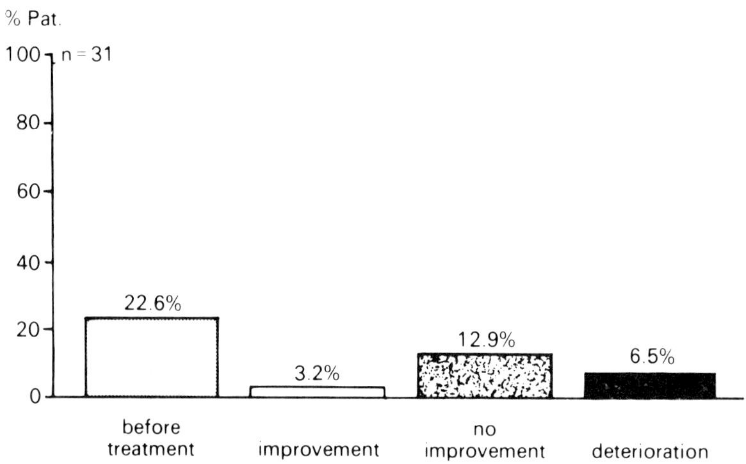

Figure 4 Premenstrual breast tenderness and the effect of therapy

Eight patients (27%) suffered serious premenstrual headaches. Four improved with therapy, three showed no change, and one got worse (Figure 3).

Seven patients (23%) complained of premenstrual breast tenderness. Only one improved after 3 months of therapy. Four showed no change, and two got worse (Figure 4).

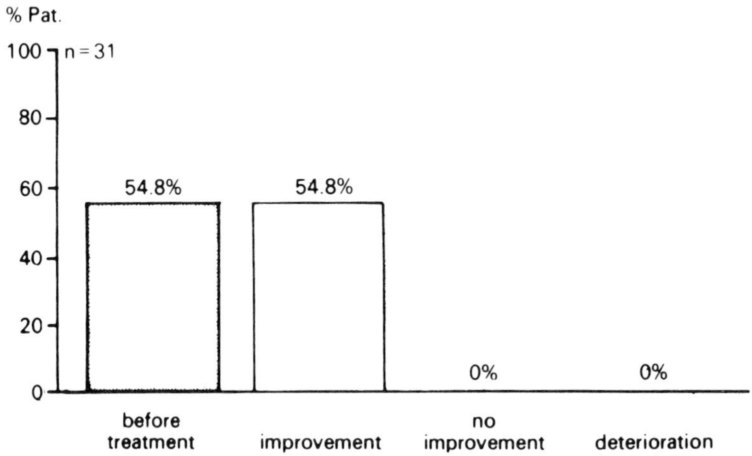

Figure 5 Hypermenorrhoea and the effect of dydrogesterone (20 mg/day in the second half of the menstrual cycle, days 15 to 25)

In addition to various premenstrual problems, 17 of the women complained of hypermenorrhoea and 13 suffered dysmenorrhoea, the two problems sometimes both occurring in the same patient. The administration of dydrogesterone for these patients' premenstrual complaints proved a very effective therapy for these problems too. All 17 of the women suffering hypermenorrhoea reported that this problem completely disappeared upon treatment with dydrogesterone (Figure 5). This was also the case in 11/13 (85%) of the women suffering dysmenorrhoea (Figure 6).

Prior to treatment, 22 of the patients in this study group

suffered irregular cycles with occasional spottings, or abnormally long cycles, most of which were anovulatory. Serum progesterone levels were low in these patients and basal body

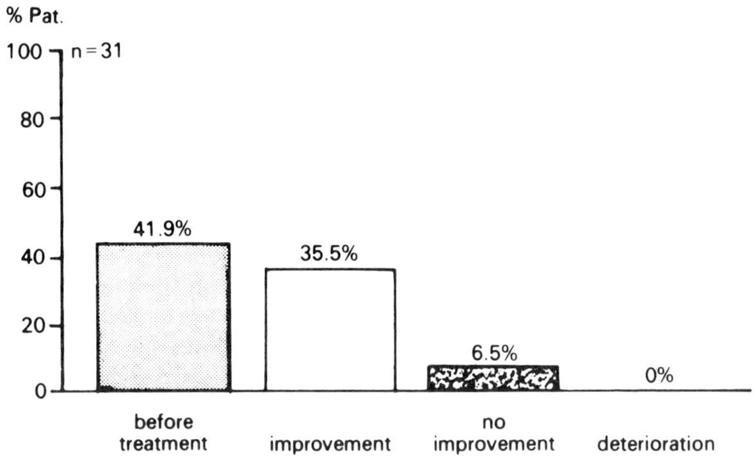

Figure 6 Dysmenorrhoea and the effect of dydrogesterone (20 mg/day in the second half of the menstrual cycle, days 15 to 25)

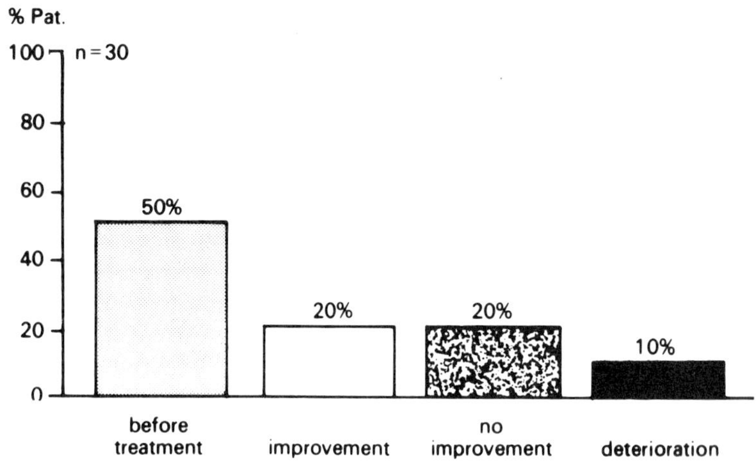

Figure 7 Effect of 20 mg/day dydrogesterone on diminished libido

temperature recordings monophasic. Treatment with dydrogesterone for these patients' premenstrual problems resulted in the 'normalization' of the cycle in 20 cases. There was no improvement in one patient and in another bleeding irregularities became worse. (As dydrogesterone does not affect basal body temperature, it may be safely given to sterility patients without fear that it will cause temperature changes which might erroneously be taken to indicate that ovulation has taken place.)

A high percentage (50%) of the women taking part in this investigation complained prior to treatment that they were experiencing diminished libido. Treatment with dydrogesterone corrected this problem in six of the women (40%).

No side-effects were seen during the 3-month course of treatment. In particular it should be mentioned that no adverse effects were seen on the gastrointestinal tract, and that no anabolic or virilizing effects occurred.

DISCUSSION

Although conducted in a small number of women, this study leads to the conclusion that patients suffering premenstrual problems are likely to find dydrogesterone of benefit in alleviating individual symptoms. In particular, it can be expected to have good results on premenstrual depression, headaches and oedema.

The study further shows dydrogesterone to be effective in the treatment of dysmenorrhoea and hypermenorrhoea, and in 'normalizing' the menstrual cycle.

It is not our intention on this occasion to make a detailed comparison of the effects of one progestational compound versus those of another, but we would just briefly mention that results similar to those given above have also been seen after the administration of the 17α-hydroxyprogesterone derivative, medroxyprogesterone acetate (Farlutal®). With a dosage of

10 mg/day, good results were seen on premenstrual oedema, depression and headaches, and also on dysmenorrhoea, hypermenorrhoea and spotting, but again the effect on premenstrual breast tenderness was limited (Table 1).

To some extent medroxyprogesterone acetate was not quite as well tolerated as dydrogesterone, some patients complaining of weariness, of lethargy, and of mild depression, though other patients, it will be remembered, found their feelings of depression alleviated by this substance.

Table 1 Effects on premenstrual symptoms of 10 mg/day medroxyprogesterone acetate administered during the second half of the premenstrual cycle. (Figures indicate number of patients)

	Before treatment	*Improvement*	*No change*	*Deterioration*
Breast tenderness	6	1	4	2
Depression	6	3	1	2
Dysmenorrhoea	9	6	3	0
Headache	7	3	2	2
Hypermenorrhoea	13	7	4	2
Oedema	5	3	1	1
Spotting	11	9	2	0

The generally good results reported with progestogen therapy in the second half of the menstrual cycle support the theory that a deficiency of progesterone is an important factor in the aetiology of the premenstrual syndrome. It may well be that an insufficient progesterone production by the corpus luteum during this phase of the cycle, resulting perhaps from an elevated level of prolactin and prostaglandins, contributes to, if not causes, the premenstrual syndrome. It is known that prolactin levels rapidly rise in any stress situation, and it is possible that even the mildest of psychological stresses in the days prior to menstruation is sufficient to cause a substantial rise in the prolactin level.

This paper is dedicated to Professor K. Knörr on the occasion of his 65th birthday.

References

Benedek-Jaszmann, L. J. and Hearn-Sturtevant, M. D. (1976). Premenstrual tension and functional infertility. *Lancet*, **1**, 1095

Litschgi, M. and Glatthaar, E. (1978). Primäre Dysmenorrhö und Hyperprolaktinämie. *Geburtsh. u. Frauenheilk.*, **38**, 569

Timonen, S. and Procopé, B. J. (1973). The premenstrual syndrome; frequency and association of symptoms. *Ann. Chir. Gynaecol. Fenn.*, **62**, 108

6
A double-blind, placebo-controlled, multi-centre study of the efficacy of dydrogesterone (Duphaston®)

A. A. HASPELS

INTRODUCTION

The publication in recent years of a considerable number of articles on the subject of the premenstrual syndrome (PMS), both in the scientific and in the lay press, has led to an increased interest in this subject and to an improvement in our knowledge about it.

The term 'PMS' covers a wide range of psychic and somatic symptoms which occur during the post-ovulatory phase of the menstrual cycle. The most frequently mentioned are: irritability, depression, exhaustion, panic, aggression, painful or tender breasts, oedema, stomach ache, weight gain, nausea and headache.

In the past PMS was treated in a variety of ways, no single therapy emerging as the most generally accepted one.

In the late seventies, however, Taylor (1977) reported that dydrogesterone could be regarded as a safe and effective therapy for general use. In order to further examine the efficacy of this drug, a large double-blind, placebo-controlled, multi-centre study was recently conducted in Belgium, France, the United

Kingdom and The Netherlands. The data reported in the present paper are those collected in the Dutch part of this study.

MATERIALS AND METHODS

The study protocol called for the investigators to report on the efficacy of dydrogesterone and to compare this drug with placebo in women suffering at least three complaints generally regarded as being part of the premenstrual syndrome. These complaints had to occur regularly in the second half of the menstrual cycle and to disappear as soon as the menstrual bleeding began. The women had to be between 18 and 40 years of age, and the study group was to exclude women suffering from any other gynaecological disturbances, those using oral contraceptives, and those who had suffered any psychiatric illness during the two preceding years.

As laid down in the protocol, the study covered four consecutive menstrual cycles. The first, when no treatment was given, was regarded as a baseline cycle, during which information was collected about the pre-treatment PMS complaints of each patient. In the three subsequent cycles the patients were treated on a double-blind basis with dydrogesterone (Duphaston[R]) or indistinguishable placebo. Counting the first day of a menstrual bleeding as day 1, treatment began on day 12 of each cycle and continued until the first day of the next cycle. The dosage was one 10 mg tablet twice per day, i.e. 20 mg per day.

The patients were asked to carefully record the occurrence of any of the following 14 symptoms:

(a) *Psychic symptoms*
Aggression, depression, drowsiness, exhaustion, fainting, irritability and panic.

(b) *Somatic symptoms*
Backache, headache, nausea, oedema, painful or tender breasts, stomach ache and weight gain.

They were also asked to rate each complaint using the scale: 0 = non-existent, 1 = light, 2 = moderate, 3 = severe.

The figures relating to symptoms which were clearly pre-menstrual ones, in that they occurred only during the 2 weeks before menstruation, were added together to form the patient's

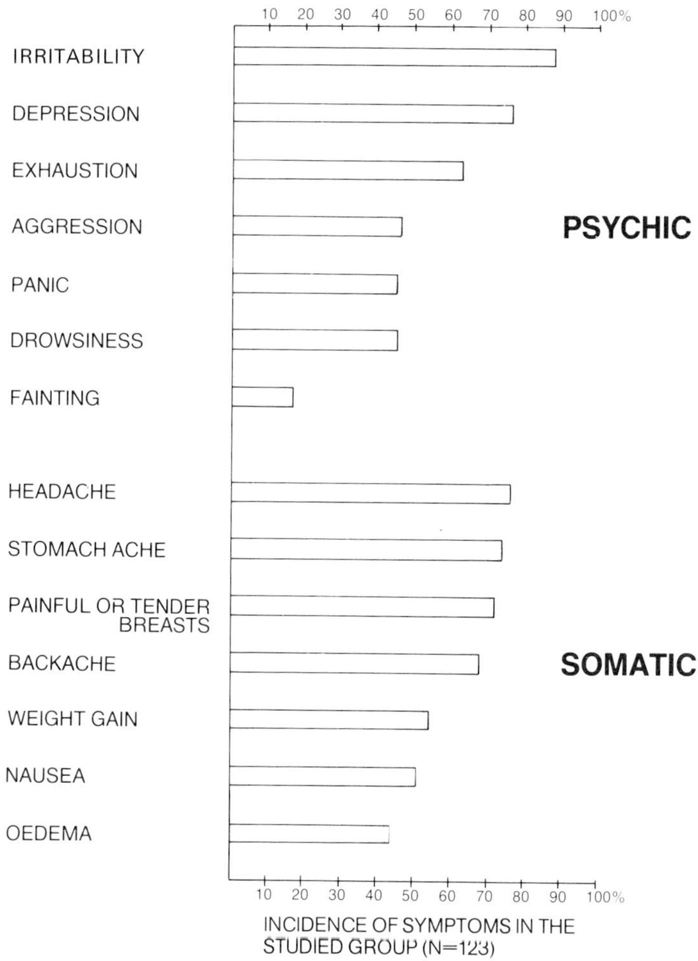

Figure 1 Incidence of PMS symptoms before treatment

'score'. The score of each patient in each cycle was thus between 0 (no symptoms at all) and 42 (the severe occurrence of all 14 symptoms).

The investigators recorded the presence and severity of the symptoms during the baseline period and again during the third treated cycle, and the difference between the two scores was taken as an indication of the efficacy of the treatment.

Sixty investigators collaborated in the Dutch study: five gynaecologists and 55 general practitioners.

Initially 150 PMS sufferers agreed to take part in this project; agreed, that is, to undergo treatment and to keep a daily diary of their symptoms for 4 months. Of these 150, 27 subsequently dropped out. The most frequently given reason was that it was difficult to comply with the request to record symptoms each day, but three patients dropped out after two treated cycles because of a lack of improvement in their symptoms. The remaining 123 women completed their course of treatment, and it is the data of these cases which form the basis of this report.

When the study was completed it was found that of the original 150 patients, 83 had received dydrogesterone, whilst 67 had received placebo. Of the 123 cases eventually available for analysis, 70 had received the active substance, whilst 53 had been given placebo.

Statistical analysis was done by means of the χ^2-test (two-sided) or Fisher test (two-sided).

RESULTS

The symptoms experienced by the 123 patients during the pre-treatment baseline menstrual cycle were:

(a) *Psychic symptoms*
Irritability 89%, depression 76%, exhaustion 63%, aggression 46%, drowsiness 45%, panic 45%, fainting 17%.

(b) *Somatic symptoms*
Headache 76%, stomach ache 74%, painful or tender breasts 72%, backache 68%, weight gain 54%, nausea 51%, oedema 44% (Figure 1).

The change in the patients' PMS after 3 months of treatment was assessed by comparing each patient's post-treatment score with her pre-treatment score, using the previously mentioned scale of 0 = non-existent, 1 = light, 2 = moderate, 3 = severe.

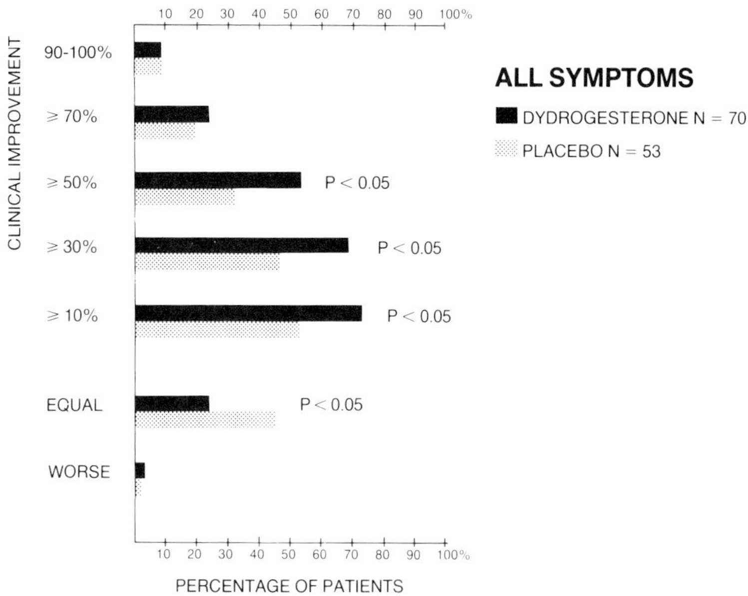

Figure 2 Degree of change in PMS after 3 months' treatment with 20 mg/day dydrogesterone or with placebo

A post-treatment score four or more points lower than the pre-treatment score indicated an improvement in the patient's PMS, a score differing by up to four points either way indicated that no change had occurred, and a score which was four or more points higher indicated that the patient's PMS had become worse. This calculation showed that an improvement had occurred in 73% of the women treated with dydrogesterone, against 53% of those in the placebo group, a statistically significant difference in favour of dydrogesterone ($p<0.05$) (see Figure 2).

The degree of change by the third month of treatment was assessed by expressing the difference between each patient's pre- and post-treatment score as a percentage of the pre-treatment figure.

$$\text{Improvement} = \frac{\text{number of points improved}}{\text{pre-treatment score}} \times 100\%$$

This calculation revealed:

	dydro-gesterone ($n = 70$)	placebo ($n = 53$)	p (χ^2)
50–100% improvement	54%	32%	0.02
10–49% improvement	19%	21%	NS
Unchanged	24%	45%	0.02
Worsened	3%	2%	NS

As can be seen from Figure 2, the number of cases in which the PMS became worse during the treatment period was very small; it amounted to 3% in the dydrogesterone group and to 2% in the placebo group.

Using the same notation scheme, 0 = non-existent, 1 = light, 2 = moderate and 3 = severe, the effects of therapy on individual symptoms after 3 months' therapy could also be calculated. These changes could be grouped as:

(a) disappeared completely, i.e. moves from $3 \rightarrow 0$, $2 \rightarrow 0$ or $1 \rightarrow 0$;
(b) improved considerably, a move of two points from $3 \rightarrow 1$;
(c) improved slightly, a move of one point, from $3 \rightarrow 2$, or from $2 \rightarrow 1$;
(d) no change;
(e) a worsening.

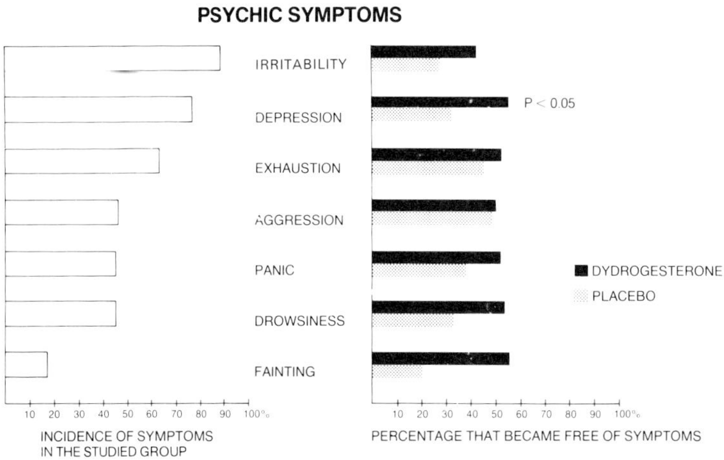

Figure 3 Percentages of psychic PMS symptoms which disappeared completely upon 3 months' treatment with 20 mg/day dydrogesterone or with placebo

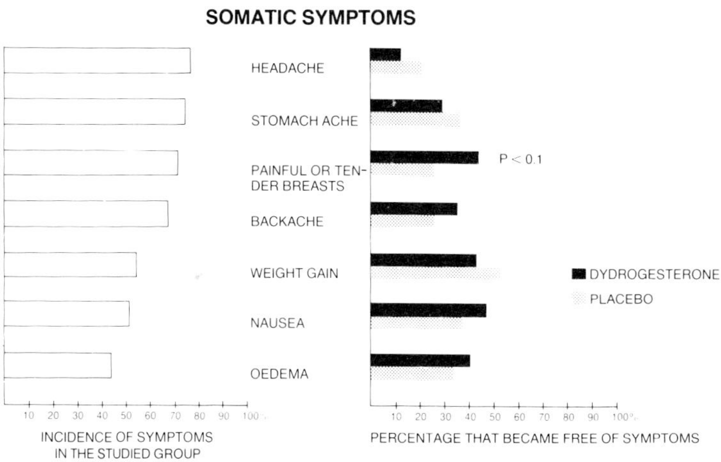

Figure 4 Percentages of somatic PMS symptoms which disappeared completely upon 3 months' treatment with 20 mg/day dydrogesterone or with placebo

A number of women in both treatment groups reported that individual symptoms had disappeared completely by the third month of treatment, but the effectiveness of dydrogesterone was greater than that of placebo in this respect, particularly in connection with the psychic symptoms (Table 1, Figures 3 and 4). Dydrogesterone succeeded in removing individual psychic

Table 1 PMS symptoms which disappeared completely upon 3 months' treatment with 20 mg/day dydrogesterone or with placebo

	Dydrogesterone (n = 70)		*Placebo (n = 53)*		*P (Two-sided χ^2-test)*
	Total	*Symptom-free*	*Total*	*Symptom-free*	
Psychic symptoms					
Irritability	64	27	44	12	0.17
Depression	55	30	38	12	0.048
Exhaustion	48	25	29	13	0.70
Aggression	32	16	25	12	0.91
Panic	31	16	24	9	0.44
Drowsiness	34	18	21	7	0.25
Fainting	11	6	10	2	0.23*
Somatic symptoms					
Headache	49	6	45	9	0.46
Stomach ache	49	14	42	15	0.61
Painful or tender breasts	51	22	38	10	0.07
Backache	46	16	38	10	0.39
Weight gain	37	16	29	15	0.66
Nausea	36	17	27	10	0.58
Oedema	30	12	24	8	0.83

* Fisher test

symptoms in at least 42% of all instances. A statistically significant difference between the two therapy groups, in favour of the active drug, emerged in this respect in connection with depression ($p < 0.05$), and a nearly statistically significant difference emerged for painful and tender breasts ($p < 0.1$).

Table 2 Percentages of moderate and severe symptoms which were cured or considerably improved, and those which were slightly improved, by 3 months' treatment with 20 mg/day dydrogesterone or placebo

	Psychic symptoms		*Somatic symptoms*	
	Dydro-gesterone	*Placebo*	*Dydro-gesterone*	*Placebo*
Cured or considerably improved symptoms (a move of two points)	55%	35%	37%	20%
Slightly improved symptoms (a move of one point)	18%	23%	26%	23%
Average number of moderate and severe symptoms per patient	2.6	2.6	3.1	3.0

A study of the *moderate* and *severe* symptoms which had disappeared or improved considerably and of those which had improved slightly, again showed the superiority of dydrogesterone, particularly in connection with the psychic symptoms (Table 2).

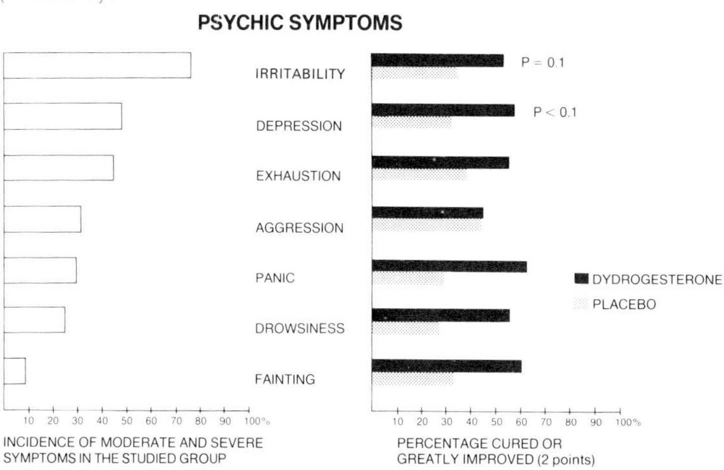

Figure 5 Percentages of moderate or severe psychic PMS symptoms which were cured or considerably improved by 3 months' treatment with 20 mg/day dydrogesterone or with placebo

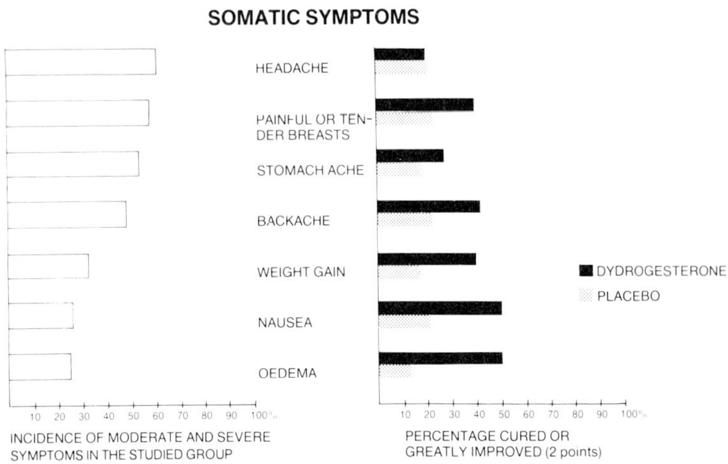

Figure 6 Percentages of moderate or severe somatic PMS symptoms which were cured or considerably improved by 3 months' treatment with 20 mg/day dydrogesterone or with placebo

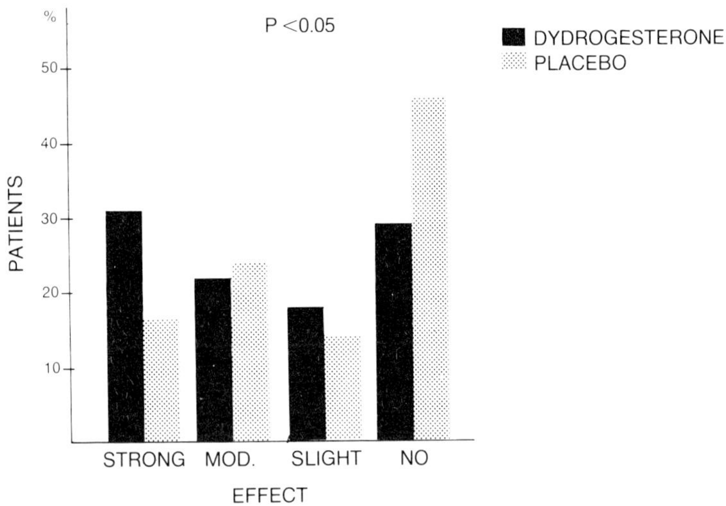

Figure 7 Patients' opinions on the efficacy of the treatment they received for their PMS problems

The number of moderate and severe symptoms which disappeared completely or which improved considerably are detailed in Table 3 and illustrated in Figures 5 and 6. The superiority of dydrogesterone over placebo reached the nearly statistically significant level in connection with irritability ($p = 0.1$) and with depression ($p < 0.1$).

Table 3 Severe or moderate PMS symptoms which were cured or considerably improved by 3 months' treatment with 20 mg/day dydrogesterone or placebo

	Dydrogesterone (n = 70)		*Placebo (n = 53)*		*P (Two-sided χ^2-test)*
	Total	*Cured or considerably improved*	*Total*	*Cured or considerably improved*	
Psychic symptoms					
Irritability	53	28	40	14	0.12
Depression	31	18	28	9	0.07
Exhaustion	33	18	21	8	0.46
Aggression	20	9	18	8	0.84
Panic	21	13	14	4	0.16
Drowsiness	20	11	11	3	0.37*
Fainting	5	3	6	2	0.78*
Somatic symptoms					
Headache	41	8	34	7	0.96
Painful or tender breasts	41	16	31	7	0.26
Stomach ache	33	9	33	6	0.60
Backache	33	14	27	6	0.30
Weight gain	29	13	12	2	0.23*
Nausea	18	9	14	3	0.25
Oedema	22	11	8	1	0.15*

* Fisher test

Finally, the 123 women who completed the 3-month course of treatment were asked how effective they themselves thought their treatment had been. When answering they could choose between: very effective, moderately effective, slightly effective and not effective. The replies of the 70 women treated with dydrogesterone were: very effective 31%, moderately effective 22%, slightly effective 18% and not effective 29%. The figures in the placebo group were 16%, 24%, 14% and 46% respectively. When subjected to analysis by the Wilcoxon test, this resulted in a statistically significant difference in favour of dydrogesterone ($p < 0.05$) (Figure 7).

CONCLUSIONS

On the basis of the study reported here, it is concluded that:

1. Dydrogesterone is a statistically significant better therapy than placebo for the psychic symptoms of the premenstrual syndrome.
2. It is clinically better than placebo for the treatment of the somatic PMS symptoms.
3. Moderate and severe PMS symptoms respond far better to dydrogesterone than to placebo.

Reference

Taylor, R. W. (1977). The treatment of premenstrual tension with dydrogesterone ('Duphaston'). *Curr. Med. Res. Opin.*, **4** (Suppl. 4), 35

Discussion A – between panel members

Chairman: W. H. UTIAN

Utian: I have listened very carefully to your presentations and am struck by the fact that most of you who gave a definition of the syndrome seemed to describe it in rather different terms. Does this mean in practice that you accept everyone coming to you who says that she is suffering from premenstrual symptoms as being a patient with PMS?

Dennerstein: In general at the clinic in Melbourne we accept all women for treatment who label themselves as suffering from PMS, but we define as PMS patients ourselves only those who have symptoms 7 to 10 days premenstrually and who do not have those symptoms during the rest of the cycle. Of course everyone has ups-and-downs, but we only include major changes in our list of symptoms.

Utian: So it would be correct to say that we are dealing with a syndrome consisting of a variety of symptoms that occur cyclically 7 to 10 days before menstruation and that appear to be related to ovulatory cycles, although the degree of the luteal phase involvement is in question. There appears to be a wide variety of symptoms and a wide variety in the degree of severity, but there are few, if any, specific symptoms. Is it therefore the general way in which the patient presents herself and her problem which allows us to attach the label 'PMS'?

Strecker: I am quite willing to accept the definition you have given, but does the premenstrual syndrome occur in ovulatory cycles only or does it also occur in anovulatory cycles? We have checked our patients for this and found that whilst most cycles in which the premenstrual syndrome occurs are ovulatory, there are a few exceptions.

van Keep: In our epidemiological data we found that women taking oral contraceptive pills also sometimes have premenstrual problems, albeit a bit less than in women not taking oral contraceptives. Women with an IUD, by the way, appear as a group to have more complaints than other women.

Utian: Let us now turn to the aetiology of the premenstrual syndrome. The introductory papers have mentioned a number of interesting factors that may be related to the aetiology of the syndrome. In the last 5 years we have learned that quite a number of symptoms and syndromes, some of which were previously described as being purely behavioural in causation, are tying in more and more in psycho-neuro-endocrinology. You have talked about altered hormone level ratios, but what really is leading to these altered ratios? Does anyone have information on the causal relationships beyond the altered oestrogen and progesterone ratio? What is happening at the brain level? Is anything known about the endorphin mechanism in this connection? What happens to hypothalamic functions which affect the ovarian function and, hence, the steroid profiles that have been mentioned?

Day: If I have to answer this question, I am bound to say that not much is known about all this. Professor Butt in Birmingham has done some studies in a small number of patients, about 12, looking in particular at the releasing hormone, FSH, LH and the resulting steroid profiles. As far as I know, nothing is known about the endorphin mechanism in these patients and the consequences that the changed endorphin mechanism might have.

We shall certainly learn more about this in the future. Assays are already available in some centres.

Utian: Well, the question of assays is a very difficult one, because what is measured in peripheral samples may be quite different from what happens centrally in the neuro-peptidergic pathways.

van Keep: Any effort to explain the premenstrual syndrome through endorphin changes seems to me to be pure speculation. There is a tendency nowadays to hold endorphins responsible for anything and everything that we cannot explain in other ways, and in my opinion this goes too far. If we accept a major role of the neuro-hormones in the aetiology of the premenstrual syndrome, there is a danger that we shall be tempted not to look for the best possible therapy with the means that are available now.

Haspels: In view of the high percentage of patients who react well to placebo therapy, I am still inclined to believe that there is a possibility that endorphins do play a role in the aetiology of the syndrome.

Utian: So, anything that we might be inclined to say further on this subject would, for the moment, be pure speculation. When you try to measure hormones there are so many variables. The levels may vary according to the time of the day – or night – that they are measured. They may be related to the level of stress the woman is exposed to. This goes particularly for prolactin, but also for gonadotrophins. Have patients with a premenstrual syndrome been brought into hospital and observed over a period of several days, also during their sleep? Have studies been done in which blood was drawn at regular intervals? Have the patients' endocrine profiles really been evaluated in depth?

Day: Basically the answer to your question is 'no'. No such in-depth studies have been done. It would not be very easy to persuade a patient with a premenstrual syndrome to be hospitalized for 2 or 3 weeks, and I personally am not sure that it would be justifiable to do so.

Strecker: The prolactin levels in PMS patients are usually within normal limits, but there may be pulsatile peaks in the second half of the cycle, and these peaks probably inhibit the LH production and thus the corpus luteum function. Consequently, the progesterone levels may be a little lower than normal.

Dennerstein: It was in view of the fact that hormone levels can fluctuate so much and be influenced by so many factors that we chose to do hormone assays from timed urine collections. It is a crude method, but it is an easy one. When we find an abnormal cycle we continue into the next cycle. We also continue if a woman says that she has had an unusual month as far as stress is concerned.

Strecker: I should like, if I may, to give my ideas about the aetiology of the premenstrual syndrome. One main reason for the occurrence of PMS seems to be an imbalance of the oestrogen:progesterone ratio. This is mostly caused by low serum progesterone concentrations during the second half of the menstrual cycle, which result in an insufficient corpus luteum function. As Timonen and Procopé have shown (*Ann. Chir. Gynaecol. Fenn.*, 1973, **62**, 108), a decrease of oestrogens in addition to low progesterone values increases the sensitivity of the myometrium to prostaglandins and oxytocin. This can certainly cause the symptom of uterine cramps. On the other hand, aldosterone that can be responsible for oedema and fluid retention is normally suppressed by progesterone (Taubert and Kuhl, *Med. Mschr.*, 1974, **28**, 27). Aldosterone is often found to be increased in women suffering PMS. Another important factor in the aetiology of PMS is hyperprolactinaemia. Elevated concentrations of serum prolactin cause corpus luteum insufficiency; moreover, they have a direct influence on the water and electrolyte balance (Litschgi and Glatthaar, *Geburtsh. u. Frauenheilk.*, 1978, **38**, 569). Hyperprolactinaemia quite often occurs in conjunction with elevated prostaglandin concentrations (Litschgi and Glatthaar, 1978). As I mentioned in my paper, prolactin is a hormone which is dependent on environmental factors and, especially, on the stress situations of everyday life. A certain psychic tension is easily able to produce peaks of prolactin concentrations which join the causative chain leading to PMS.

Utian: To some extent it seems to me that we are dealing with a syndrome which presents us with a 'chicken or the egg' situation. There may be a syndrome and we may find changed endocrine profiles, but we do not know which comes first. Let us now move away from the hypothetical background and devote a little time to clinical matters. I should like to ask the clinicians present how they arrive at their diagnosis? Dr Dennerstein has said that she accepts initially anyone who labels herself as being a premenstrual syndrome patient, but what, I wonder, makes you exclude a patient from this possible diagnosis?

Dennerstein: I make sure at the first interview that those who have major psychiatric disorders are excluded – those who are suffering from depressive illness, schizophrenia and anxiety states, or who have done so in the past. I also exclude after the first interview those whose doctors have labelled them as suffering from premenstrual tension but who are well aware themselves that this is not the case. To give you an example of this: I recently saw a patient whose doctor sent her in because he was certain that she had a premenstrual syndrome. At the first interview the patient told me that her problem was that she could not decide whether to continue with an affair or not, and that because of this she was becoming increasingly depressed. As this, unfortunately, happened to coincide with the premenstrual phase, her doctor diagnosed the depression as being part of a premenstrual syndrome. She couldn't tell her doctor about the affair because she was worried that he, being a male, might tell her husband. This is the type of complication that we see rather often and it is for this reason that a lengthy and good first contact is needed in order to be sure that the problem we are dealing with really is premenstrual syndrome.

Sampson: I agree with Dr Dennerstein. As I run a clinic which is quite specifically for premenstrual syndrome patients, the label has already been given to the women I see, either by the patient herself or by the doctor who has referred her to the clinic. The first thing one has to do is to take a very good history and exclude patients with psychiatric illnesses. One also has to exclude gross gynaecological abnormalities. I don't make a diagnosis at a first interview unless the case is very clear indeed. I explain to the patient what I think the premenstrual syndrome really is. Some women then immediately say 'That is not what I have got; I am in the wrong place'. Others may say that they think that's what they've got. I then give the patient a chart on which to record all symptoms for at least one menstrual cycle. We then study the chart together. I don't feed these data into the computer at a next interview, because it is usually quite easy to see from the chart when the symptoms occur. If, after this first cycle, both the patient and I are convinced that we are dealing with a premenstrual syndrome, we begin to think about therapy. If we are not sure, the patient will be asked to keep a chart for one more cycle.

Utian: Who should treat the patient then? Should it be the practising gynaecologist? Should it be an endocrinologist? A physician? A psychiatrist? Or should it simply be the person to whom the patient first presents herself? What should this person do before he can say: now I have the diagnosis and now we can start to think about treatment?

Dennerstein: This is really a good point. The situations that Dr Sampson and I have described are not really typical ones.

We have enough manpower and time and expertise and are also engaged in research. It is another thing to tell people what they should do in practice. For the majority of patients one does not need to spend very much time on hormonal evaluation. The most important thing is that one is prepared to sit and listen and accept the woman's symptoms. The doctor should then help the patient to gain insight into what is happening. Further help might not be needed. With regard to your question 'who should treat the PMS patient?' I think it should be the doctor she first calls upon.

Sampson: I have a slightly different view here. One should make a thorough assessment before any therapy is begun. This also goes for general practitioners. Most women keep records of their menstrual cycles, even if it is only the ringing of the days of their menses in a diary. They do so for all sorts of reasons. It is easy for women also to keep a record of their premenstrual difficulties. These records enable one to be sure that a problem is a premenstrual one before one begins to treat it. I really do think this very important.

Day: I have my doubts here. If you are going to give everybody a chart and ask them to keep detailed records of all their problems you risk having the general practitioners inundated with patients with all sorts of symptoms, minor ones as well as major ones. You could, in a way, attract far more than the genuine PMS sufferer.

van Keep: In our study in France we asked women if they had ever sought medical advice for their premenstrual problems. Despite the fact that so many of the women taking part in the survey had at one time or another suffered such problems (77% of the 2500 respondents were still experiencing them), less than half had ever sought medical help in this connection. Incidentally, only 38% of the 795 who had consulted a doctor had arranged to see him specifically in connection with their premenstrual problems; the others had discussed the matter with the doctor when visiting him primarily in another connection. It was interesting to see what women said when we asked them *why* they had not consulted their doctors. A high percentage, 31%, said that they had not done so because the premenstrual syndrome is a 'natural phenomenon', 44% said that their problems did not warrant it, 9% said that it was pointless to seek medical advice for such problems, 6% said that they did not like taking medicaments, and 5% said that they preferred their own remedies. Finally, somewhat strangely, 5% said that they 'would not dare' to consult a doctor about such matters.

Utian: I think that these data are interesting because in general our opinions and knowledge are based only on the women who do come and ask our advice. It is quite unusual to know the thoughts of those who do not come to consult us. I have a few more questions relating to the clinical evaluation. What does the clinician do when he sees such a patient? Is there, apart from checking the chart of symptoms, anything in particular he should do? Should he ask the woman to keep a basal body temperature record? Are there any specific blood tests to be done, or would this be a waste of time and money?

Sampson: We have heard already that most women with premenstrual syndrome have ovulatory cycles, but also that some women on the pill have PMS problems. A basal body temperature chart would not therefore be very useful. As far as blood sampling is concerned, I cannot see what conclusions could be drawn from this which would really be of benefit to the patient. Personally I avoid blood sampling in the clinic, partly because I have not found it to provide answers for the patients and partly because it does not provide answers to my research questions either.

Strecker: It is part of my routine to check whether the patient ovulates or not, but for me the basal body temperature is not the best way to find this out. I prefer to check progesterone levels in the second half of the cycle. Before this is done, however, a very careful analysis is made of the medical history of the patient.

Dennerstein: There are some patients with very unusual symptoms and I believe that in some of these cases a daily hormonal evaluation may well prove useful. Fluctuations in hormone levels can then be compared with symptoms. I have now, for example, two patients who lose their voice each month premenstrually. I would be interested to hear if anybody else has ever seen patients with this problem. You can imagine that we are most interested to see what happens endocrinologically to these patients. Hormonal profiles are also interesting in patients with menstrual migraine.

Utian: Let us now move on to another aspect: the treatment of the patient. We have heard about treatment with progesterone. We know that progesterone can be administered via different routes, and we have heard about oral progestogens and their effects. But we should also discuss bromocriptine, pyridoxine, diuretics, ovulation inhibitors, and so on. Perhaps you would give me your opinions on these different regimens?

Haspels: In my presentation earlier I told you of my experience with the progestogen, dydrogesterone. I have no experience with pyridoxine in the premenstrual syndrome. I do know, however, from my experience with the contraceptive pill and with oestrogen therapy in the post-menopause, that there may be a change in tryptophan metabolism as a result of oestrogen administration. For this reason we routinely recommend the administration of pyridoxine, 125 mg/day in the tablet-free days. My clinical impression is that this is effective. The situation might well not be the same when one is talking about premenstrual depression. If a woman is depressed it is indeed possible that this is due to a pyridoxine deficiency, and, hence, to a disturbed tryptophan metabolism, but it is most likely that depression from this cause would occur throughout the cycle, not just during the premenstrual phase. I also have the impression from the literature that any effects of pyridoxine on PMS are very close to those of placebo therapy.

Day: I can confirm that last point. In a study carried out in 70 PMS patients we found similar improvement rates with both substances.

van Keep: Jaszmann has advocated the use of bromocriptine in cases in which PMS is associated with increased prolactin levels. I think he may have a point, but we seem to have agreed that PMS should be treated by the general practitioner, and I would personally hesitate before allowing a powerful drug like bromocriptine out of the hands of a specialist.

Strecker: I agree. Also, it is rather difficult to bring increased prolactin levels to normal with bromocriptine treatment, since if one suppresses prolactin levels too far, there will also be an insufficient corpus luteum. I personally think it might be better, and less complicated, to correct the oestrogen/progesterone imbalance in PMS patients by augmenting low progesterone levels through the administration of oral progestogens.

Utian: And now a provocative question: if there is evidence that placebo therapy is in many instances as good as any other therapy, why don't we just give placebo to all patients? Why don't we restrict ourselves to the administration of TLC (tender loving care)?

Dennerstein: I would not go as far as that, but I do know that if you listen to a patient's complaints, if you explain to her what is happening and help her to accept it, you will certainly find that such psychological therapies can be extremely helpful in quite a number of cases.

Haspels: I will, of course, not deny the value of these psychologically supporting techniques, but since we all believe that an endocrine disorder is the basis of this syndrome, I am personally inclined to believe that hormone therapy might in the end be superior. I am, however, quite willing to accept that a combination of both therapies might give the best results of all.

van Keep: If it is, as our exercise with the French data tends to suggest, that there are three or more syndromes beneath the present PMS umbrella, it is likely that we shall need three or more different therapies in order to satisfactorily handle all PMS problems. The lack of adequate diagnostic techniques for PMS means that at the present time there is little alternative but to try various therapies and see how the patient reacts to them. Although some of the more powerful drugs mentioned in this discussion should clearly be used only in special situations, others, dydrogesterone for example, seem very safe and effective.

Day: I am inclined to agree with that. We have had very good results with dydrogesterone, and definitely found it to be superior to placebo. We also use bromocriptine and danazol, but would not favour these drugs being handled by the general practitioner. Some patients, it is true, may need specific anti-depressive therapy, but these may well be patients who are suffering another problem, and not PMS as such.

Utian: I think we have now covered sufficient aspects of premenstrual syndrome in this discussion between the panel members. I thank you all for your contributions and I now return the chair to Dr van Keep for a discussion in which members of the audience are invited to join.

Discussion B – between panel members and members of the audience

Chairman: P. A. VAN KEEP

van Keep: The first discussion session was devoted to the putting straight of a number of very basic points, and it was, I thought, a very well-structured discussion. The present discussion can take whichever direction you, the members of the audience, wish. We might, nevertheless, start from the same point. Are there any comments on the panel members' opinions on terminology?

K. Dalton (London, UK): There was one point which was perhaps not as strongly made as it might have been, and that is that in order to qualify as being part of the premenstrual syndrome, a complaint, besides being present during the second half of the menstrual cycle, has to be completely *absent* for at least 7 days post-menstrually. A complicating factor, and an error to my mind, arises when one uses the Moos Menstrual Distress Questionnaire in this connection. This questionnaire contains a list of symptoms, a very long list, which includes not only premenstrual symptoms but also, as its name implies, other problems which lead to menstrual distress, i.e. endometriosis, dyspareunia, chronic depression, problems arising from ovarian cysts, salpingitis, etc., problems which exist throughout the month but which get *worse* premenstrually. In my opinion, and this is particularly important when one is considering treatment, if one is talking about premenstrual syndrome, one must discard all these problems and consider solely those which exist only during the premenstrual phase, and not during the rest of the cycle. My definition of the premenstrual syndrome, therefore, is symptoms which occur regularly in the 2 weeks before menstruation, but not at all for at least 1 week after menstruation.

van Keep: Thank you. I think that is a fair comment.

L. J. Jaszmann (Bennekom, The Netherlands): The point I should like to make is a small one really, but I should like to draw attention to the fact that premenstrual symptoms sometimes occur even when there is no menstruation: in cases where a hysterectomy but not an oophorectomy has been done for instance.

Strecker: In that case I would consider the term 'premenstrual' inappropriate. A group of investigators in Belgium recently found that after hysterectomy the blood supply to the ovaries may be partially impaired (Donnez *et al.*, 1980, *Acta Endocrinol. (Kbh)*, Suppl. 234, 106). The result is an insufficient function of the corpus luteum, which might explain why these patients suffer from a 'premenstrual' syndrome.

van Keep: Would you not then expect climacteric complaints rather than PMS ones? Richards has described this phenomenon in about 70% of hysterectomized patients (*Lancet*, 1974, **2**, 983).

Strecker: The patients I mentioned were ovulating, they had a biphasic temperature curve but low progesterone levels.

Utian: The findings that Dr Strecker mentioned are in disagreement with our data from the University of Cape Town and the Case Western Reserve University in Cleveland. We looked very closely at ovarian function before, during and following hysterectomy, studying pituitary hormones and steroid profiles. We found that, although there are very interesting changes in the immediate post-operative phase, within a short space of time patients go back to full normal ovulatory function.

van Keep: Whether or not the blood supply of the ovaries – and consequently also ovarian function – is affected by hysterectomy might depend on the surgical skill of the gynaecologist.

Dalton: There may also be differences according to why the hysterectomy was done in the first place. I have found, however, that the patient who has premenstrual syndrome before hysterectomy often has the same symptoms after the operation.

van Keep: So much for hysterectomy. Let us return to the more frequently seen situation, that of the woman with complaints preceding menstruation.

M. de Senarclens (Geneva, Switzerland): What I have missed in everything that was said is that we should try to understand the significance that such complaints have for the individual patient. If you begin to administer hormones before you have any real insight into the underlying problems you might make mistakes. Secondly, it might be interesting to see how the hormonal profiles of patients develop under psychotherapy alone.

Dennerstein: I agree with you. I personally believe that we should try very hard to understand what the symptoms mean for the individual patient. This is why I insist on a detailed assessment being made at the outset. Some people find that their complaints become tolerable once they understand what is happening. As far as your second point is concerned, a study of the hormonal correlates of psychotherapy would indeed be interesting, and it would not be a difficult thing to do.

de Senarclens: How many patients drop out during treatment? In my experience PMS patients have a tendency to change doctors. I have 'inherited' quite a number from other gynaecologists. The premenstrual syndrome does seem to be one of the most stubborn syndromes in gynaecology.

Sampson: My problem is that the patients do not drop out but keep coming to the clinic!

Dennerstein: We do not have many drop-outs either. In the beginning of our work it was different. In the early days I attempted to explain to people that their problems were largely psychologically based. Some people could not cope with this and dropped out, or changed doctors, because they really wanted a medical explanation. We are now more careful in our explanation. We explain that they are people who are, for some reason, biologically unstable during this, premenstrual, phase of their cycle; sometimes we can see from their hormonal profiles why this is the case, but in most cases we cannot.

van Keep: Perhaps we could now move on to therapy.

Dalton: I do believe in a special place for progesterone (the natural substance) – as opposed to progestogen (the synthetic) – therapy. There are progesterone receptors in the brain at the site of the hypothalamus and there is no evidence that there is any affinity between these receptors and progestogen. Progesterone has a short half-life in blood, but if you administer it by injection you may see a rise in blood levels that lasts up to 48 hours. Given in a suppository, the rise is shorter; it is never longer than 24 hours and sometimes as short as 6 or 12 hours. This means that, using this method, you have to give progesterone twice or three times a day. If you give a progestogen – whichever one – you lower the blood progesterone levels, and that makes quite a difference.

Haspels: Yet I think that there is good evidence from double-blind studies that progestogens such as dydrogesterone are very effective. I have a question for Dr Dalton. You have found a good reaction both to placebo therapy and to therapy with progesterone, but what do you do in the case of patients who do not respond to either therapy?

Dalton: As I use a very precise definition for premenstrual syndrome and treat only patients with what one might call a 'pure' premenstrual syndrome, I find initially a very high response to natural progesterone. If patients do not react to a low dose I increase the dose until they do. My success rate is near to 90%. The dose is an individually tailored one, and what I am aiming at is a long-term satisfactory result.

de Senarclens: What is the effect on the psyche of hyperprolactinaemia?

Strecker: Many patients with prolactinomas – and consequently with high levels of serum prolactin – do not show any psychic abnormality. They may have amenorrhoea, a fertility problem, retain body fluids and electrolytes, but otherwise they feel fine. It is possible, however, that the endocrine findings are the consequence of a psychic problem, because stress increases the prolactin levels, as I mentioned in my paper.

Utian: The recent apparent escalation in prolactinoma is probably due to our increased diagnostic ability. There is another possibility, however. Stress leads to hyperprolactinaemia, via stimulation of the lactotropes. From this it could be concluded that an increase in stress in the population could feasibly lead to an increase in the incidence of prolactinoma.

van Keep: Let us look at the impact of the premenstrual syndrome on the woman's daily life.

M. Smith (Perth, Australia): Many women seek help because the premenstrual syndrome is destroying their family life, their behaviour being so bad for 1 or 2 weeks out of every 4. Should the husband not be routinely interviewed in these cases? Should we not talk to the husband and, in so doing, offer him some help so that he can support his wife? There is often a great gulf between the two partners. Often neither the woman nor her husband really understands why the woman is behaving as she is.

Dennerstein: I am glad that you have raised this point because the premenstrual syndrome really can have devastating effects on the family. I know of at least three divorces because of the hostility and aversion which existed during the premenstrual phase. I have at this moment a most difficult case in which the wife suffers a very severe premenstrual and menstrual migraine. Her psychological profile is essentially normal, which in view of the stress she is under is rather surprising. Each month she is completely incapacitated and bed-ridden for 6 days. Her two young children have to go to her in-laws, and her husband has to take time off from work to nurse her. We do see the husbands of premenstrual patients. Often we do not have to invite them because they come spontaneously with their wives, so glad are they that something might finally be done about the problem. Also, they want to tell us how bad it is for them. They also want, however, to know what they can do to help. When you ask women what their husbands have done to help them in the past, they often say that what their husbands did was not very helpful and in fact even made it worse. When you succeed in sorting out what is helpful for the individual woman you are really helping the couple.

Day: I have a similar experience. I sometimes feel too that PMS can be a manifestation of minor problems in a couple's relationship.

van Keep: Now to a somewhat wider aspect: the influence of culture as a whole on the syndrome. We could not find differences between socio-economic groups, but what about completely different cultures? Does the syndrome occur, for example, in primitive cultures?

Smith: I spent a number of years working in obstetrics and gynaecology in New Guinea. The difference between the European-type people and the local people there in this respect was very clear. As far as the local people were concerned, I never saw people presenting with premenstrual tension nor with climacteric complaints, but in the European-type people I certainly did. This may be explained by the whole cultural situation and by the differing attitudes towards menstruation. There might perhaps also be another explanation: because of the number of pregnancies which more primitive people tend to have, and more especially, because of their prolonged periods of nursing, women in these societies do not have many menstrual cycles. The question then is: what is normal, their situation or ours? It may be that, in the general scheme of things, the quick succession of menstrual cycles is abnormal and that this, as such, causes stress. There is no doubt in my mind that this stress is the major factor in the aetiology of the syndrome. We have to find out what women feel about menstruation and what in these feelings leads to the stress that induces the endocrine changes.

van Keep: Hardly any studies have been done into this, and those that have been done have mostly been undertaken in rather biased groups: medical students, psychology students, nurses, etc. Even less is known about attitudes towards menstruation in primitive societies. I wonder, however, if you are not taking it too far when you are seeking an explanation of the premenstrual syndrome in women's attitudes towards the menstrual bleeding.

A. H. Crisp (London, UK): Those who postulate psychosocial stress as a causal factor in premenstrual syndrome seem to imply that this stress occurs only in the second half of the menstrual cycle. Is this right? If so, what exactly are the stresses which occur at this time?

Sampson: I personally consider premenstrual syndrome to be a psychosomatic disorder, using the definition of Linford Rees; that is to say, the basic changes, in this case the hormonal ones, are physiological, and it is these which in a way alter the patient's vulnerability. The stresses are there all the time; it is the vulnerability which is the intrinsic factor.

Dennerstein: Attributional-type studies have demonstrated that the expectations of both men and women affect their behaviour; that is to say, if people expect to feel badly they *will* feel badly. In premenstrual syndrome, the fact that some women expect to feel badly is loaded on to the fact that they are biologically unstable at that time. They are, therefore, more likely to suffer when stresses occur at that time than when the same stresses occur during the follicular phase of the cycle or even during the ovulatory phase.

J. Laferla (Ann Arbor, USA): I am impressed by the multi-dimensional approach of this workshop. I think this approach is particularly important when one is looking at the aetiology of premenstrual syndrome. I do not think that the showing of altered progesterone levels or of the fact that progestogens eliminate premenstrual complaints, in isolation, give us much clue as to the aetiology. We really have to look at this problem in a multi-factorial way. Following on from this, I would just mention that some research which I have been doing recently may have a bearing on this subject, albeit perhaps tangentially. When looking at sexual arousal in men, we found that when coming into the laboratory for the first time to see sexually stimulating films, many subjects had a low FSH. When we administered various sorts of multi-emotion scales, the only factor which was different was the anxiety one. There is, then, a rather clear association between high anxiety and low FSH. It may be that studies looking at the effects of stress could well be useful in the study of premenstrual syndrome. One last point, and this is in connection with the comments about the effect of premenstrual syndrome on family life. A colleague of mine in Rochester, New York, has recently made a study of alcoholism in young women, and found a very striking correlation between young female alcoholism and premenstrual distress. I am not suggesting that premenstrual syndrome causes alcoholism, but it may be that there are factors at work which influence both.

van Keep: Jaszmann and I once presumed that the process of symptom formation in the premenstrual syndrome, in dysmenorrhoea and in climacteric complaints must take place along similar lines; in all three instances there is a somatic problem and a psychic problem, and we presumed that these two potentiated one another. However, when we looked at women over 50 and recorded the incidence of earlier premenstrual problems and dysmenorrhoea, and of climacteric complaints, we found that whilst there was indeed a strong correlation between the premenstrual syndrome and dysmenorrhoea, there was none at all between these and climacteric complaints. We concluded from this that the mechanisms of symptom formation must be quite different (van Keep and Jaszmann, *Geburtsh. u. Frauenheilk.*, 1973, **8,** 669).

Dennerstein: We too found, using psychological rating scales, that PMS patients and climacteric patients are very different. The patients with climacteric problems were within the normal range of the Eysenck Personality Inventory ratings in neuroticism, extraversion, anxiety and depression, whereas the PMS patients were certainly not.

Dalton: In my experience women with premenstrual syndrome tend to have less climacteric problems. My explanation for this is that women with a premenstrual syndrome have a high oestrogen production during their menstruating years, so high that it upsets the oestrogen:progesterone ratio and leads to PMS problems. However, when these women approach the menopause their high oestrogen levels protect them to a large extent from climacteric problems.

van Keep: You might also consider this, rather sexist and not too seriously offered, explanation: to some extent women might suffer premenstrual problems because they resent having menstruation, and, by the same token, those who resent losing their fertility may suffer climacteric problems. However, whichever explanation might be correct, the hormonal one or the psychological one, we found no relation at all between the two syndromes.

Strecker: I am coming back to Dr Laferla's remark on low FSH and anxiety being correlated in men: if women are suffering from stress in the *first* half of their cycle this will lead to an increase in serum prolactin levels, which, in turn, will lead to decrease of gonadotrophins. This low FSH and LH might impair follicle growth and corpus luteum formation, and this could indeed lead to premenstrual problems.

van Keep: We shall have to stop this discussion soon. Is there a last question please?

Utian: Since you brought the matter up yourself, I should like to ask whether we are in fact dealing with one syndrome or with several.

Haspels: Apart from the evidence presented by Dr van Keep, I feel that the diversity of the complaints and the different reactions of patients to therapy suggest that we may indeed be dealing with different syndromes. It would be very helpful if these syndromes could be identified.

Dennerstein: I certainly know that women respond variously to treatment. I do not know whether or not I have to conclude from this that there are different syndromes, but it certainly strengthens my view that there should be an individual assessment and approach to management in each case.

Day: I would obviously like it if the complaints would break down into a few clusters, each with a different aetiology and requiring different treatment, but I am not sure that the data presented provide us with the answer.

Sampson: My work shows that certain symptoms do respond differently. I could, therefore, support the theory that there are different syndromes, but this subject needs further study.

Strecker: The term 'syndrome' is often used when a group of individual symptoms is ill-defined and vague.

van Keep: For me a 'syndrome' is a group of symptoms with a common cause. When reporting our findings and presenting the theory of there being several syndromes instead of just one, we entitled our paper 'A statistical exercise'. The clusters of symptoms were proposed not by the statistical analysis but by ourselves. What the subsequent analysis did was to show that our thoughts, based on clinical experience, were indeed logical. Considering the costs involved, a repetition of this exercise including endocrine profiles as explaining variables is not really justified. Nevertheless, the concept resulting from our exercise is attractive because it may explain why some women react to a given therapy whilst others do not. It is clear that further work is needed before one will be able to say exactly which symptoms of the premenstrual syndrome in which women will react to which therapies. At present we have to find the best therapy for an individual patient by trial and error, though our knowledge, particularly through well-controlled clinical studies, is increasing all the time. Common sense and, above all, an open mind for our individual patient's problems and needs, remain important factors in the handling of the PMS patient.

Thank you all for your contributions to this workshop.

Index